Black holes are the most fascinating discovery of modern astronomy. They have already become legendary, and form the basis of many myths and fantasies. Are they really the monsters of science fiction which devour light and stars? Are they purely hypothetical objects from the theory of relativity or are they an observable reality?

In answering these questions, the author takes us on a fabulous voyage through space and time. He explains how stars are born, light up and die. He takes us into the strange world of supernovae, X-ray stars and quasars. We travel on a journey to the very edge of the Universe and to the limits of contemporary physics.

Dr Jean-Pierre Luminet is a French specialist in black holes. He is an astronomer at the Meudon Observatory, conducting research in relativistic astrophysics and cosmology for the Centre National de la Recherche Scientifique (CNRS). In recognition of his work, he was awarded the prize of the Société Astronomique de France and the medal of the CNRS. Also a writer, he has published various essays and collections of poems, and he is now working on a series of screenplays on great men in science.

Black holes

Black Holes

Jean-Pierre Luminet
Observatoire de Paris-Meudon

Translated from the French
by Alison Bullough and Andrew King

Published by the Press Syndicate of the University of Cambridge
The Pitt Building, Trumpington Street, Cambridge CB2 1RP
40 West 20th Street, New York, NY 10011–4211, USA
10 Stamford Road, Oakleigh, Melbourne 3166, Australia

Originally published in French as *Les Trous Noirs*
by Belfond, Paris, 1987 and © Belfond 1987

First published in English, with minor revisions,
by Cambridge University Press 1992 as *Black Holes*
English edition © Cambridge University Press 1992
Reprinted 1993

Printed in Great Britain at the University Press, Cambridge

A catalogue record for this book is available from the British Library

Library of Congress cataloguing in publication data

Luminet, Jean-Pierre.
[Trous noirs. English]
Black holes / Jean-Pierre Luminet: translated by Alison Bullough and Andrew King
p. cm.
Translation of: Les trous noirs.
Includes bibliographical references and index.
ISBN 0-521-40029-5. – ISBN 0-521-40906-3 (pbk.)
1. Black holes (Astronomy) I. Title.
QB843.B55L8613 1992
523.8'875–dc20 91–27238 CIP

ISBN 0 521 40029 5 hardback
ISBN 0 521 40906 3 paperback

WD

I dedicate this book to all those people for whom each answer is another question.

'Books are not made to be believed, but to be subjected to inquiry. When we consider a book, we must not ask ourselves what is says but what is means . . . '

Umberto Eco, *The Name of the Rose*

Contents

Foreword to the French edition

The history of the Universe and its evolution brings us again and again to a basic examination of our own history and of the understanding of our environment. Astrophysicists have the formidable privilege of having the largest view of the Universe; particle detectors and large telescopes are today used to study distant stars, and throughout space and time, from the infinitely large to the infinitely small, the Universe never ceases to surprise us by revealing its structures little by little.

In this respect, we live in a very exciting time. Every year, there is a major event in astronomy: 1986 was marked by the exploration of Uranus and Halley's comet, which increased our understanding of the origins of the Solar System; 1989 saw the exploration of Neptune, 4.5 billion kilometres away from us; 1990 the launch of the Space Telescope. . . And 1987? That year will be long remembered as the year of the supernova in the Magellanic Cloud. Astronomers had been waiting for this rare and shortlived event for 4 centuries, and it was made even more exceptional because it may have produced 'before our very eyes' the strangest star in the Universe: a black hole.

The 'invention' of black holes was undoubtedly one of the most daring intellectual feats of modern times. Even the words 'black hole' have a magical connotation: there are stars in the Universe which one cannot see, but which are able to absorb all the surrounding matter; closed worlds, completely cut off from our own, opening onto bottomless pits in which everything, all matter, is inexorably sucked and crushed. . . Black holes have such strange properties that for a long time they were hardly deemed credible, although they were very popular with sensationalists. Our understanding of space and time is probed to its depths by black holes. However, their existence has been whole-heartedly embraced by the general public and they have formed the centrepiece for many science fiction stories, comic strips and disaster movies.

Although somewhat mysterious, black holes are much less 'magical' than more familiar speculations such as the existence of life around stars other than the Sun. My friend Jean-Pierre Luminet, who is able to understand and formulate the most complex notions in the theories of gravity and relativity, and to share his discoveries by relating them in simple terms, has written a comprehensive work on the nature and existence of black holes. In this book he takes us on a voyage through space and time where the forces of gravity make the space-time continuum submit to their every whim. At the end of this carefully mapped cosmic journey, the reader will understand that the concept of a black hole is not so mysterious, and that there really exist stars in the sky which have all the characteristics of black holes. The author shows us that black holes are a powerful analytical tool, a sort of theoretical telescope with variable magnification, able to examine macroscopic and microscopic mysteries, thus providing us with one of the fundamental keys to our understanding of the Universe.

Black holes began as an intellectual adventure, and perhaps in the future they will symbolise adventure itself in the eyes of cosmic travellers who, like the sailors of antiquity, will think of them as the edge of the World which will swallow them into oblivion and the unknown. Finally, I hope that, like me, the reader of this book will feel that the developments in modern astrophysics are as fascinating to discover and digest as the most fantastic science fiction novels.

Jean Audouze

Foreword to the English edition

Black holes are the bane of modern astronomy. Invisible by definition, their existence has proved difficult to substantiate, yet black holes have captured the popular imagination in a way that no other astronomical objects have succeeded in doing. Black holes are time machines, as well as openings to other universes. These concepts are more than science fiction. In a lucid exposition, Jean-Pierre Luminet makes the many aspects of black holes accessible to the non-experts. This book on black hole physics translates the esoteric concepts of what black holes are and how they may be detected into a readable discourse.

The author is a remarkable man: poet, musician, artist and scientist. Much of this diverse background can be discerned via his elegant metaphors and colourful descriptions. The book excels, however, in the explanations of the theoretical background to black holes, descriptions of their astronomical counterparts and, above all, of the exotic phenomena that make black holes such objects of mystery. The reader will learn how black holes form, and the fate of space travellers who venture near black holes. They are utilisable as energy machines, and may constitute the final destiny of the Universe, to which all matter must eventually succumb.

Astronomers believe that giant black holes lurk at the centre of many galaxies. Occasionally the orbit of a star takes it dangerously close, resulting in disruption of the star and emission of an intense burst of radiation as the stellar debris falls at ever increasing speed into the black hole. The feeding of these monsters is believed to occur in the dense cores of galaxies, where these objects first formed aeons ago. More modest black holes, of stellar mass, are the endpoints of massive stars which terminate their lives as an exploding supernova when the stellar core collapses into a black hole. Very small black holes are conjectured to form near the beginning of the Universe, and these mini black holes can disrupt themselves in an intense explosion. There is no escape from a black hole, yet in-falling matter vanishes from our view most probably to end up in a separate universe.

All of these phenomena, and many more, are described by Jean Pierre Luminet without compromising the underlying physics, yet at a level that manages to be eminently readable. This book will elucidate many of the topics covered in Stephen Hawking's *A Brief History of Time*: it is a recommend sequel. If you wish to understand black holes, or white holes, or even wormholes, this book will provide enlightenment. The reader will undertake a fascinating and enjoyable journey through space and time that covers many of the novel ideas and discoveries at the frontier of astronomy.

Joseph Silk
University of California at Berkeley

Acknowledgements

Jean Audouze, Philippe André, René Lachal, Jean-Alain Marck, Sylvano Bonazzola, Christian Poinas, Brandon Carter, Laurent Nottale, Andrew King and Egidio Landi kindly but ruthlessly read, discussed and corrected my manuscript. I am profoundly grateful to them.

PART 1
GRAVITATION AND LIGHT

'Theories are like fishing: it is only by casting into unknown waters that you may catch something.'

Novalis

1

First fruits

The happiest of mortals

'The weight of a small bird is sufficient to move the Earth.'
Leonardo da Vinci

The Greek philosophers, still unsurpassed in many domains of intellectual thought, had very little understanding of gravitation. Aristotle believed that each body had its 'natural place' in the Universe. At the bottom was the Earth and all that was directly attached to it, while above this lay water, then air and finally the lightest element of all, fire. A body subjected to forces dislodging it from its natural place moved in such a way as to return to its own level. Thus an arrow or stone thrown into the air fell back to the ground in an effort to return to its natural place at the centre of the Earth. Moreover, Aristotle asserted that this motion was rectilinear; he believed that an arrow shot from a bow would travel upwards in a straight line; when the force provided by the bow ceased to act, it would fall back to Earth, also in a straight line.

The theories developed by the Greek philosophers remained almost unquestioned for twenty centuries, despite daily evidence to the contrary; in reality an arrow follows a curved not straight trajectory, a parabolic arc. Only John Philoponus, who lived in Alexandria in the sixth century, dared to challenge Aristotelian thinking in suggesting the principle of inertia.

Galileo was the first person to subject gravity to proper scientific scrutiny. He conducted a series of experiments which involved dropping all manner of objects from the leaning tower of Pisa, and rolling different sized balls down inclined planes. In 1638, he discovered the most fundamental property of gravity: under it, *all bodies experience the same acceleration, regardless of their mass or chemical composition.*

Galileo's work was distinguished by his careful observation of physical phenomena and his abstract scientific reasoning. His work was in clear opposition to Aristotle's understanding of the World. To study a physical phenomenon, we have to separate out all the extraneous effects which complicate our everyday experience of it. To deduce the universal law governing the free fall of bodies in a *vacuum* from observations made on objects falling through air, Galileo first of all had to understand the forces of friction and air resistance. These are 'parasitic' phenomena whose effects depend on the size and mass of the object involved and which mask the real action of gravity.[1]

The intuitive genius was followed by the analytical genius. According to popular legend, on a night of full moon in 1666 Isaac Newton was sitting thinking under an apple tree, when an apple fell off. It suddenly occurred to him that the Moon and the apple might both be falling towards the Earth because of the same attractive force, the Earth's gravitation. He calculated that the force of attraction between two bodies decreased as the inverse square of their distance; by doubling the distance between two bodies, the attractive force between them is reduced by four. Since the Moon was 60 times further away from the centre of the Earth than the apple,[2] it was therefore falling with an acceleration of $60 \times 60 = 3600$ times smaller than the apple's. He then applied Galileo's law concerning the free fall of a body; that the distance fallen was proportional to the acceleration, and to the square of the time. Newton thus deduced that the fruit took one second to fall the same distance as the Moon fell in one minute. The Moon's motion was well known,

[1] If, as some historians assert, Galileo did not in fact throw objects off the leaning tower of Pisa, he deserves even greater credit for producing his theory from a chain of abstract reasoning!

[2] 384 000 kilometres for the Moon as opposed to 6400 kilometres for the apple.

and Newton's estimated distances agreed. He had just discovered the law of universal attraction.

Newton's work (which extended much further than his theory of gravity) had an enormous effect on contemporary thought, and embodies some of the most glorious feats of human intelligence. A century later, Pierre Simon Laplace, the French 'father' of black holes, recognised in Newton's *Mathematical Principles of Natural Philosophy* 'its pre-eminence over other products of the human spirit'. The mathematician Joseph Lagrange went even further: 'as there is only one Universe to explain, no-one can repeat what was done by Newton, the happiest of mortals'. Founding a scientific theory may or may not lead to personal happiness; but there was certainly no scientific work of such fundamental importance until our view of space and time was radically remodelled by Albert Einstein.

The appetite of the planets

The most spectacular application of Newton's theory is in celestial mechanics. Newton used his law of universal attraction ('universal' because *all bodies* are subject to gravity) to explain Kepler's empirical laws describing 'the appetite of the planets' for the Sun. Armed with this astonishingly accurate tool scientists began intoxicatedly to measure out the new Solar System.

The first triumph of this new mechanics was Edmund Halley's prediction of the return of 'his' comet in 1759. The comet actually reappeared on Christmas day 1758.

In addition, Newton's theory revealed that Kepler's description of planetary motion was only approximate. If each planet were attracted only by the Sun, its orbit would be a perfect ellipse. However, each planet is perturbed by the attraction of the others (especially Jupiter, by far the largest planet). These deviations, although small, can be calculated and are observable; they are the subject of the powerful 'theory of perturbations' used by Urbain Le Verrier and John Adams to predict the existence and position of a new planet in 1846. The discovery of *Neptune* at exactly the calculated position marked the high point of the Newtonian theory of gravitation.

Two forerunners of the invisible worlds

'There exist in the heavens therefore dark bodies, as large as and perhaps as numerous as the stars themselves. Rays from a luminous star having the same density as the Earth and a diameter 250 times that of the Sun would not reach us because of its gravitational attraction; it is therefore possible that the largest luminous bodies in the Universe may be invisible for this reason.'

Pierre Simon Laplace (1796)

At the end of the eighteenth century, the Reverend John Michell and Pierre Simon Laplace combined the idea of a *finite* velocity of light with Newton's concept of an *escape velocity*, and discovered the most fascinating consequence of gravitational attraction: the *black hole.*

The concept of escape velocity is very familiar. We know that a stone thrown into the air returns to the ground, regardless of the force with which it is launched. We recognise the effect of inexorably attracting gravity. However, we should ask up to what point gravity imprisons matter. What holds for a small stone thrown upwards from the Earth already fails for a stone thrown upwards from one of Mars' small moons, such as Phobos. On Phobos the force of gravity is so small that a man's arm would be strong enough to put a small stone *into orbit* around it, or even send it into orbit about Mars itself, some 9000 kilometres away.

Let us return to the Earth. Terrestrial gravity can be represented as a deep well which widens out at the top. A projectile can only escape from it if its velocity is sufficiently high. To put a satellite into orbit, the rocket launcher has to reach a certain altitude, incline itself to be parallel to the Earth's surface and then increase its velocity to at least 8 km/s. At this velocity the centrifugal force (directed into space) balances the gravitational force (directed towards the centre of the Earth).

At fairgrounds one sometimes sees a track (the wall of death), with very steeply banked walls; motorcyclists rise up the walls as their velocities increase. A satellite moving in an orbit has simply stabilised itself on the walls of the gravitational well.

If the motorcyclist increases his velocity above a certain critical value, he flies off the top of the banking. In the same way, if the rocket has a sufficiently high velocity, it will be able to overcome the Earth's gravitational attraction. This critical velocity – identical for a pebble or a rocket – is called the *escape velocity*. On the Earth's surface it is 11.2 km/s, and it can easily be calculated for any planet, star or other celestial body. The nature of the projectile is irrelevant: the escape velocity depends only on the global properties of the star from which it is to be launched. The greater the mass, the greater the escape velocity, and for a given mass the escape velocity will increase as the radius of the star decreases.

In other words, the more dense or *compact* a star is, the deeper its gravitational well and the harder it is for objects to escape its influence, as is intuitively obvious. The escape velocity is only 5 m/s for Phobos and 2.4 km/s for the Moon, but 620 km/s for the Sun. From a more dense star, such as a *white dwarf* (see Chapter 5), it reaches several thousand kilometres per second.

The notion of a black hole ultimately derives from the simple concept of escape velocity. The velocity of light has been known to be about 300 000 km/s since Olaüs Roemer's observations of Jupiter's moons in 1676. It is easy to imagine the existence of stars so massive that the escape velocity from their surface is greater than the velocity of light.

In an article read to the Royal Society in 1783 and published later in *Philosophical Transactions*, John Michell wrote: 'If the semi-diameter of a sphere of the same density with the sun were to exceed that of the sun in proportion of 500 to 1, a body falling from an infinite height towards it, would have acquired at its surface a greater velocity than light, and consequently, supposing light to be attracted by the same force in proportion to its *vis inertiae*, with other bodies, all light emitted from such a body would be made to return towards it, by its own proper gravity.' A little later, in 1796, the mathematician and astronomer Pierre Simon, Marquis de Laplace, the prince of celestial mechanics, made similar remarks in his *Exposition du système du monde*.

Besides being more than a century before their time in imagining light being trapped by gravity, Laplace and Michell envisaged enormous dark bodies that could be as numerous as the stars. At the end of the twentieth century, so rich in scientific upheavals, the possible existence of dark matter is one of the most important questions in cosmology. It seems that a sizeable fraction of the total mass of the Universe may be hidden from view.

A gravitational theory more accurate than Newton's was needed before these invisible stars (only christened 'black holes' in 1968) could be studied in detail. Einstein's theory of General Relativity predicted the existence of black holes of the same 'size' as those guessed at by Michell and Laplace.

On closer examination, however, the agreement between Einstein's and Newton's theories for the size of these invisible stars is superficial. According to Newton, even if the escape velocity is much greater than 300 000 km/s, it is possible for light to escape from the surface of a planet up to a given altitude, before falling back (just as we can always throw a ball upwards from the Earth). But in General Relativity, it is not strictly correct to speak of escape velocity, and light cannot leave the surface of a black hole at all. Instead it stays there: the surface of a black hole is like a web of light, woven from rays which travel around the surface, never able to escape from it. We shall even see (Chapter 11) that if a black hole has its own intrinsic spin, the surface trapping the light and the surface of the black hole itself are quite distinct. Although it has an important historical and didactic value, the description of a black hole in terms of the escape velocity of light is oversimplified.

Michell's and Laplace's ideas were completely forgotten until the development of General Relativity because on the one hand there was no indication of the existence of such concentrations of matter in the Universe (for the very good reason that they were supposed to be invisible) and on the other hand, their existence rested on the hypothesis (supported by Newton) that light consisted of particles obeying the laws of gravity like ordinary matter. However, the wave theory of light, which described light as a vibration in a medium, held sway throughout the nineteenth century. According to this theory, light waves were unaffected by gravity, vitiating Michell's and Laplace's ideas.

Force fields

Planetary motion can be calculated, because we know that bodies attract each other in proportion to their masses and as the inverse square of their separation. But this leaves unanswered more profound questions such as the nature of the force of gravity, how it is generated by matter, and how it can act on bodies separated by a vacuum.

The Newtonian attractive force does not transmit itself by contact like the force of a horse pulling a cart or a gardener pushing his spade into the ground. It is generated by an object and acts *at a distance*. The idea of a force exerting itself instantaneously, without a physical medium, was quite foreign to the mechanical vision of the Universe described by René Descartes in 1644 in his *Principes de la philosophie*, which formed the basis of modern science. Newton himself, a faithful mechanicist, had prudently considered his law as a simple mathematical device to enable calculation of the motions of bodies, and not as a physical reality, writing that to imagine that gravity acted instantaneously and at a distance was an absurdity which no philosopher worthy of the name could accept. Laplace attempted to modify Newton's theory by taking into account the finite velocity of the propagation of gravity. His reasoning was sound in principle (since Einstein, we have known that gravity propagates at the velocity of light) but wrong in practice: he calculated that the velocity of the propagation of gravity must be 7 million times greater than the velocity of light.

In the nineteenth century, the same questions about action at a distance resurfaced as scientists tried to describe electricity. As with gravity, the electric force between two bodies is proportional to the product of the charges on the bodies (the gravitational force is proportional to the product of the masses), and inversely proportional to the square of their separation. But while physicists finally accepted – for want of anything better – action at a distance for gravity, they refused to accept it for electricity.

Thus Michael Faraday and James Clerk Maxwell developed the concept of a *field*, mediating the actions of bodies upon each other and propagating at finite velocity. Instead of saying that two electrical charges attract or repel each other in empty space through an instantaneous force, we can say that each of the charges produces an 'electric field' about itself whose intensity decreases with distance. The force on each charge is attributed to the *local* interaction between the two fields. The force of gravity can be described in the same way: it is exerted on all bodies in the *gravitational field* produced by another body.

This is not just a simple change in our descriptive vocabulary, the fundamental advantage of a field is that it replaces instantaneous action at a distance by action diluted in space and spread out through time. Field theory, the crowning glory of classical physics, paradoxically undermined Newtonian physics by opening up the way towards electromagnetism and then relativity.

Light according to Maxwell

At the end of the nineteenth century the forces which acted on matter could be classified into three types: gravitational, electrical and magnetic.

Electricity is characterised by the existence of two sorts of electric charge, one positive and the other negative. Charges of the same sign repel each other and charges of the opposite sign attract each other, with an intensity which varies with distance in the same way as the gravitational force. Magnetism is responsible for the behaviour of magnets, which attract iron and align themselves on the surface of the Earth in the direction of the poles. A magnet has two poles, the 'north' and the 'south': like poles repel and unlike poles attract.

In attraction and repulsion, electricity and magnetism seem to be cousins. The ancient Greeks sensed a connection. They had observed that amber (in Greek *elektron*) attracted blades of straw after it had been rubbed with a woollen scarf, and that a certain fossil resin called *magnes* attracted particles of iron. In the sixth century BC, Thales of Miletus, the most modern of Greek geometers, suggested that electricity and magnetism were manifestations of the same phenomenon and that these strange substances had a 'soul' which sucked in nearby objects.

Twenty-four centuries later, the Danish physicist, Christian Oersted, was giving a practical lesson on electricity. By chance, a magnetic needle was lying near to his equipment. Oersted noticed that every time he switched the current on, the magnetic needle was deflected. Within weeks of this discovery, André Ampère and François Arago had developed a theory describing the induction of magnetic forces by variable electric forces and vice versa. Subsequent experimental work abundantly confirmed the close relationship between electrical and magnetic phenomena.

The theory of electricity was, however, only properly justified by the experimental discovery of the *electron* in 1898. This elementary particle, a basic constituent of the atom, carries an indivisible electric charge and thus constitutes a fundamental unit of electricity. A normal atom is electrically neutral because the negative charge of the electrons surrounding the nucleus is cancelled out by the positive charges in the nucleus to which they are bound. Electric charges can be either static or in motion. For example in a metallic conductor there are free electrons which may move around. It is the motion of these charges under the influence of an electric field which gives rise to electric currents.

Similarly, magnetism in a natural magnet was found to be induced by microcurrents circulating between its molecules, and, on a much greater scale, the Earth's magnetic field is caused by enormous motions of electrically conducting matter deep in the rotating nickel-iron core. The real unification of electricity and magnetism was achieved in 1865, when Maxwell summarised all their properties in four equations: the theory of the *electromagnetic field*.

An electric charge at rest has a fixed radial field which is constant in time (Figure 1). When the charge moves, the field surrounding it has to adjust itself to the new position, and the perturbation in the field moves at a finite velocity, the velocity of light. Any displacement of the charge produces this kind of perturbation in the field; in particular, if the charge moves periodically, the perturbations in the field take the form of a wave, in the same way as a stick waggled up and down in water produces circular waves. Maxwell predicted that the periodic motion of electric charges would produce *electromagnetic waves* propagating through the vacuum at the velocity of light.

In a wave consisting of a regular pattern of crests and troughs, the

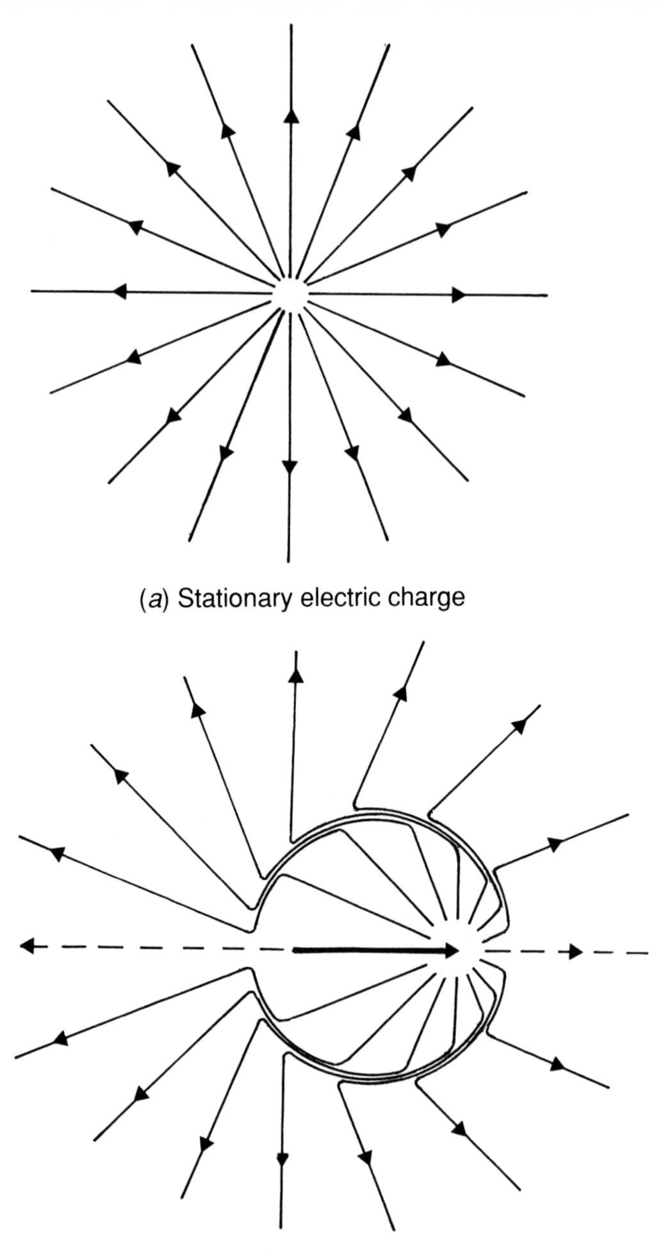

Figure 1. The electromagnetic field.

The shape of the field can be represented by lines showing the direction of the force acting on a body at a given point. (*a*) For a stationary electric charge, the field lines are radial. (*b*) When the charge moves, the perturbation in the electromagnetic field propagates outwards at the velocity of light.

Table 1. *The electromagnetic spectrum*

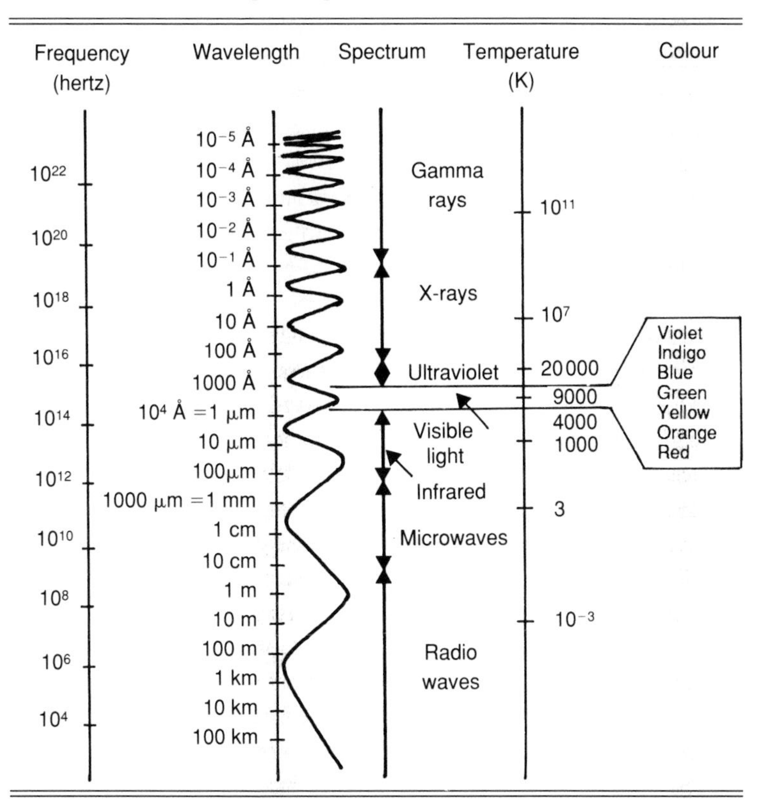

Note: Electromagnetic radiation ranges from gamma rays (the shortest wavelengths) to radio waves (the longest wavelengths). Visible light, which can be split into 'colours', represents only a minuscule part of the total spectrum. The frequency is the number of vibrations of a wave per second; it also measures the energy carried by the wave, which increases as the wavelength decreases. All bodies with temperatures above absolute zero radiate electromagnetic waves whose energy increases with temperature. The Universe has an average temperature of 2.7 degrees Kelvin and radiates microwaves; the human body emits infrared radiation (used by the military for detection); the surfaces of most of the stars at several thousand degrees Kelvin emit visible light, and those at several million degrees emit highly penetrating energetic X-rays.

distance between two consecutive crests is the *wavelength*, while the number of crests per second is the *frequency*. Visible light represents only a small part of the electromagnetic spectrum, corresponding to a narrow band of frequencies (Table 1). It is obvious that the greater the wavelength the smaller the frequency – in fact the two are inversely proportional.

Observational and theoretical astronomy are based on the properties of electromagnetic radiation. Electromagnetic waves transporting energy and momentum (higher frequencies carry more) exert a force on the matter they encounter. For example, the light illuminating this page heats and pushes against it; the electromagnetic wind radiated by the Sun is able to blow comet-tails away from it; the pressure of radiation from the core of a star can prevent the star collapsing under its own weight.

The repercussions of the electromagnetic theory were as great as those of the law of universal gravitational attraction, and had profound theoretical and practical consequences for the whole of civilisation. Maxwell died 8 years before Heinrich Hertz succeeded in producing electromagnetic waves in his laboratory (1887). At the beginning of the twentieth century Guglielmo Marconi established the first transatlantic radio transmission link. The era of telecommunications had begun.

2

Relativity

Some interference on the waves

Maxwell's theory appeared to simplify physics by unifying electricity and magnetism. In fact it complicated matters by smuggling the apple of discord into the Galilean–Newtonian description of the Universe. Detailed theoretical and experimental study of the electromagnetic field immediately raised two simple questions, which culminated in the two great theoretical scientific advances in the twentieth century: quantum mechanics and relativity.

The first question: what is the true nature of radiation? Maxwell's theory treats electromagnetic radiation as purely wave-like, but its ability to transport energy and momentum strongly suggests a particle nature. By the end of the nineteenth century a number of laboratory experiments had produced evidence of *discontinuous* radiation properties.

At the turn of the century, Max Planck put forward the hypothesis that electromagnetic waves (and light in particular) could only be radiated or absorbed as packets of energy, *quanta*. However, it was not until 1905 that Albert Einstein became the first to attribute a real existence to light quanta, now called *photons*. To explain the *photoelectric effect*, in which electrons are emitted from a metal plate irradiated by light of a sufficiently high frequency, Einstein assumed that radiation consisted of real particles carrying energy proportional to their frequency, able to knock electrons out of the metal by transferring all their energy to them. He thus revived

Newton's corpuscular theory, used by Laplace in his speculations on the trapping of light by gigantic dark stars. The apparently irreconcilable differences between mechanics and electromagnetism only truly disappeared 20 years later, when quantum mechanics took into account the wave–particle duality of all radiation and matter.

The second question: in what medium do electromagnetic waves propagate? This question probed at the very structure of space and time, in the form of the theory of relativity.

Motion and rest

The idea of *relativity* in physics, such a dazzling media success in the twentieth century, did not originate with Einstein. For three centuries, the fundamental laws of physics had been based on a principle of relativity usually attributed to Galileo, but whose correct formulation was the work of Descartes.

The use of a principle of relativity in the description of Nature indicates a sensible wish to represent physical phenomena independently of the position and motion of observers. Specifying sets of observers for whom the laws of physics appear the same amounts to determining sets of equivalent *reference frames*.

Galileo had already noticed that there was an equivalence between the observations of two experimenters, one shut in the hold of a ship, stationary with respect to the land – for instance docked in a port – and another on board a ship which was travelling away from the port in a straight line at constant velocity. If each of the experimenters dropped a ball from 1 metre above the floor of the hold, the results were identical: a vertical fall lasting 0.45 second.

Galileo had inferred from this that the ship moving away from the port had a circular motion (due to the Earth being round) and, influenced by the ancient idea of circular perfection, he had therefore deduced that the circular motion represented the 'natural' state of bodies, which was indistinguishable from rest. Descartes discovered that it was in fact the *motion of uniform translation*, that is, in an infinite straight line at a constant velocity (with neither acceleration nor deceleration), which was indistinguishable from rest. In modern times, we have all sat on a train in a station watching a neighbouring train pull slowly away, and had the impression that our own train was pulling out in the opposite direction.

These observations are simple yet profound, because they suggest that in fact *there is no difference between rest and uniform translation*. Just as rest is a state of *inertia*, a uniform translation is equivalent to rest and is also inertial.

The Principle of Inertia can be expressed in the following way: a *free* body, i.e. one not subject to any force, moves with constant velocity.

The Earth itself is an almost ideal inertial reference frame: in its orbit around the Sun, for the usual limited duration of laboratory experiments, it appears in first approximation to be moving in a straight line at a constant velocity of 30 km/s. The Earth's inertial reference frame is specified by choosing directions pointing towards fixed stars, so as to compensate for the diurnal rotation.

A marksman and a train

The Principle of Inertia gives a privileged status to the set of reference frames moving at constant velocity in that the laws of Nature take their natural form 'at rest'. Galilean relativity, and subsequently Einstein's Special Relativity, are both based on the equivalence of inertial reference frames and those moving with uniform translation.

However, it is not enough just to determine the nature of the inertial reference frames. Given a description of natural phenomena in one reference frame, a physicist must be able to describe them in any other; he needs formulae allowing him to pass from one frame to another. It is on this crucial point that Galilean relativity and Special Relativity differ.

Einstein's favourite way of illustrating these abstract notions was to take the example of a train travelling along an embankment at a constant velocity of 108 km/h, or $v = 30$ m/s. We have two inertial reference frames, the train and the embankment, the embankment representing space at rest, with respect to which the train is moving at a constant velocity. Now imagine that a man perched on the roof of one of the carriages fires a bullet in the same direction as the train is travelling. The velocity of the bullet with respect to the man is $v' = 800$ m/s.

Using the Galilean transformation formulae to transfer from the train's inertial reference frame to the embankment's inertial reference frame, the velocity of the bullet measured by an observer on the embankment is given by $v + v' = 830$ m/s. If the man turns through 180° and fires in the direction from which the train has just come, the velocity of the bullet, measured from the embankment, is $v - v' = 770$ m/s. In accord with common sense, the Galilean transformation formulae can be reduced to a simple vectorial addition of velocities.

The ether

'The ether, this child of sorrow of classical mechanics . . .'

Max Planck

If all reference frames moving with constant velocity are equivalent to one at rest, it is tempting to imagine a truly immobile frame rooted in the *absolute* space of Euclidean geometry. For Galileo, this absolute space was attached to the Sun which was the centre of the Universe. For Newton, it was the *ether*, Aristotle's fifth essence (quintessence), a perfectly rigid vibrating substance which permeated the void between physical bodies and the bodies themselves.

The advent of electromagnetic theory appeared to support the idea of the ether. It is difficult to conceive of a wave which does not have a medium through which to propagate: sound waves cause the air to vibrate, and sea waves require water. Light, an oscillation of electric and magnetic fields, must therefore be assumed also to require a vibrating medium through which to propagate independently of observers. Thus the ether could be defined as the medium through which electromagnetic waves propagate.

Let us return to the marksman on the train travelling at $v =$ 30 m/s. This time he uses a light-gun, firing photon projectiles at a velocity of 300 000 km/s. Using the Galilean transformation formulae, the observer on the embankment would measure the velocity of light to be equal to $c + v$ = 300 000.030 km/s in the same direction as the train and $c - v$ = 299 999.997 km/s in the opposite direction. Michelson and Morley's experiments, where the Earth and the ether replace the train and the embankment, show this reasoning to be false.

A rigged race

These famous experiments were carried out by Albert Michelson and Edward Morley between 1881 and 1894. They were designed to provide evidence of the Earth's absolute velocity with respect to the ether. In order to achieve this, the experimenters built a very sensitive interferometer which measured the differences between trajectories of light rays emitted in the direction of the Earth's motion and perpendicular to it. This should have enabled them to detect the absolute motion of the Earth with an accuracy of several kilometres per second.

The principle of Michelson and Morley's experiments can be described as a race between two boats moving at the same velocity c on a river flowing with a constant velocity v (Figure 2). The boats complete a return journey; boat A travels up and down the river parallel to the current, while boat B travels from one bank to the other and back. The distance travelled by each boat is exactly two river widths. By applying Pythagoras' theorem, we can see that boat B will win the race.

In Michelson and Morley's experiment, c is the velocity of light and v the velocity of the ether with respect to the Earth. However, in this race, the results are different: *the 'photon boats' always arrive at exactly the same time.* To make sense of this we have either to imagine that the Earth is completely at rest in the ether, or that the ether itself is an illusion.

With hindsight, Michelson and Morley's findings are not very surprising if we take electromagnetic theory seriously. Maxwell's theory is in evident conflict with Galilean relativity since the velocity of light appears in it as an absolute invariant, completely independent of the reference frame. The observer on the embankment will not measure the velocity of light projected by the man with the photon gun as 300 000.030 km/s or 299 999.970 km/s but as exactly 300 000 km/s, whatever the direction of the beam: the velocity of light is exactly the same in all directions and all reference frames.

Thus, since Galilean relativity was conceived to express the universality of the laws of Nature in inertial reference frames, Maxwell's equations governing electromagnetic phenomena are in flagrant contradiction with it. Galilean–Newtonian concepts of space and time are not compatible with electromagnetic theory. One or the other has to be rejected.

Special Relativity

When Einstein became aware of this conflict in 1905, he immediately decided that electromagnetism must be correct, and postulated that the velocity of light in a vacuum is an absolute constant, the maximum transmission velocity of any signal. Galilean relativity was incompatible with this postulate and had to give way to a new relativity, later called *Special Relativity*. (*General Relativity* was not formulated for another 10 years).

In passing from Galilean relativity to Special Relativity the formulae of transformation from one inertial reference frame to another are altered (in General Relativity the very nature of these frames will change). They are replaced by the 'Lorentz transformations' which conserve Maxwell's equations, and leave the velocity of light as an absolute constant.

In the experiment with the marksman on the train, the Galilean formula $w = v + v'$ for the addition of velocities is replaced by a slightly more complicated formula preserving the invariance of the velocity of light. Thus even if $v = v' = c$, w remains equal to c. At this point, the reader's common sense may rebel and assert that the observer on the embankment surely measures 830 km/s and 770 km/s, as obtained via the Galilean transformations. There is, however, no conflict, as the Lorentz formulae only differ significantly from the Galilean formulae for extremely high velocities, much greater than those normally encountered on Earth. Even if we consider the motion of the Earth about the Sun, at the already high velocity of 30 km/s, the correction required by the Lorentz formulae is only one part in ten thousand.

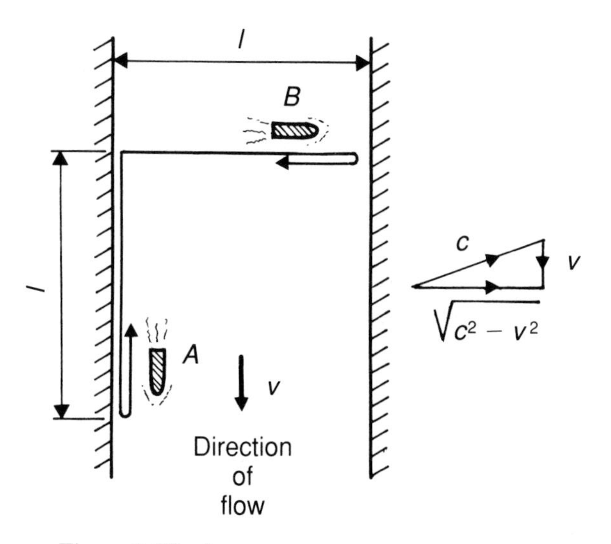

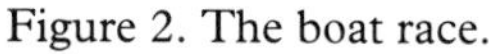

Figure 2. The boat race.

Boat A sails parallel to the current. In the first half of the race, the velocity of the current adds to the boat's own velocity c, in the second half it subtracts. The time taken to complete the course is therefore: $t_A = l/(c+v) + l/(c-v) = 2lc/(c^2 - v^2)$. Boat B sails perpendicular to the current. Its resultant velocity is given by adding the current's velocity to its own according to Pythagoras' theorem. The time taken to complete the journey is therefore $t_B = l/\sqrt{(c^2 - v^2)} + l/\sqrt{(c^2 - v^2)} = 2l/\sqrt{(c^2 - v^2)}$. Since $t_B/t_A = \sqrt{(1 - v^2/c^2)} < 1$, boat B always wins the race. (After I. Nicolson.)

Relativistic effects only become significant for velocities greater than 100 000 km/s (Table 2). Newtonian mechanics thus works perfectly well for the description of most physical phenomena, and gives correct results in all situations not involving large velocities.

A theory in the air

It takes none of the credit from Einstein to emphasise that at the turn of the century a good number of physicists were well aware of the crisis in physics brought on by Michelson and Morley's experiments. Some of them, such as Hendrik Lorentz and Henri Poincaré, gave profound insights. Lorentz was the first

(1904) to suggest the variation in time and length with the velocity of the reference frame. In 1905, Poincaré's article 'On the dynamics of the electron', published in *Comptes Rendus de l'Académie des Sciences de Paris,* introduced the mathematical formulation fully developed by Hermann Minkowski in 1908, in which time appeared as a fourth dimension. The new relativity was truly 'in the air'.

Albert Einstein's manuscript, 'On the electrodynamics of moving bodies' appeared in the German scientific publication *Annalen der Physik* a month after Poincaré's publication. It seems that Einstein, still modestly employed in the patent office in Berne, was unaware of the work of his predecessors. Special Relativity was really born because Einstein was not content just to devise formulae: he constructed a new space and time woven out of light.

The fabric of light

'The views of space and time that I wish to place before you are based on experimental physics; this is where their strength lies. They are radical. From this hour on, absolute space and time must recede to the shadows and only a kind of union of the two retain significance.'

Hermann Minkowski (1908).

In the universe of Galileo and Newton space and time were completely independent of each other. Space has three dimensions, that is, three *coordinates* are needed to fix any point in space. Space is measured by *Euclidean geometry* (geometry literally means 'measuring the Earth'). These laws are taught in school because they hold with great precision in everyday life: the shortest route from A to B is always the straight line connecting them, two parallel lines intersect only at infinity, the sum of the interior angles of a triangle is always 180°, and so on. The spatial separation of two points is independent of the observer measuring it.

Time itself is measured by a single number, but it differs from a spatial dimension by always flowing in the same direction, from the 'past' to the 'future'. This irreversible succession of phenomena

Table 2. *Relativistic effects*

$\frac{v}{c}$	Length contraction	$\frac{\text{Mass}}{\text{Rest mass}}$	Time dilation
0	1.000	1.000	1.000
0.1	0.995	1.005	0.995
0.5	0.867	1.155	0.867
0.9	0.436	2.294	0.436
0.99	0.141	7.089	0.141
0.999	0.045	22.366	0.045

The effects of Special Relativity become apparent only at velocities very close to that of light, while at low velocities, the ratios between the dimensions (length, mass and time) in motion and at rest remain close to unity.

and events, as established by observation and reason ('the cause always precedes the effect') is called *causality*.

Time, like space, is the same for all observers. Since there is no limitation of velocity, all clocks, however far apart they are in space, can be instantaneously synchronised and continue indefinitely to keep the same time. The causal structure of Galilean–Newtonian space and time can therefore be reduced to a present time extending simultaneously throughout space, separating the past and the future (Figure 3).

Space and time as absolute entities were much debated by the mathematician and philosopher Wilhelm Leibniz, a contemporary of Newton. On the basis of philosophical arguments, he maintained that space and time existed only in relation to matter. Two centuries later, Einstein's relativity partially confirmed Leibniz' view. Duration and length were effectively no longer intrinsic quantities since they depended on the velocity of the observer relative to the object measured. The Galilean–Newtonian structure of absolute space and time gave way to a new four-dimensional structure, *Minkowski's space-time.*

A point in space-time is an *event*, fixed by three space and one time coordinates. The interval between two events remains invariant (independent of the reference frame), but it is now a combination of the space and time intervals, which are no longer individually conserved.

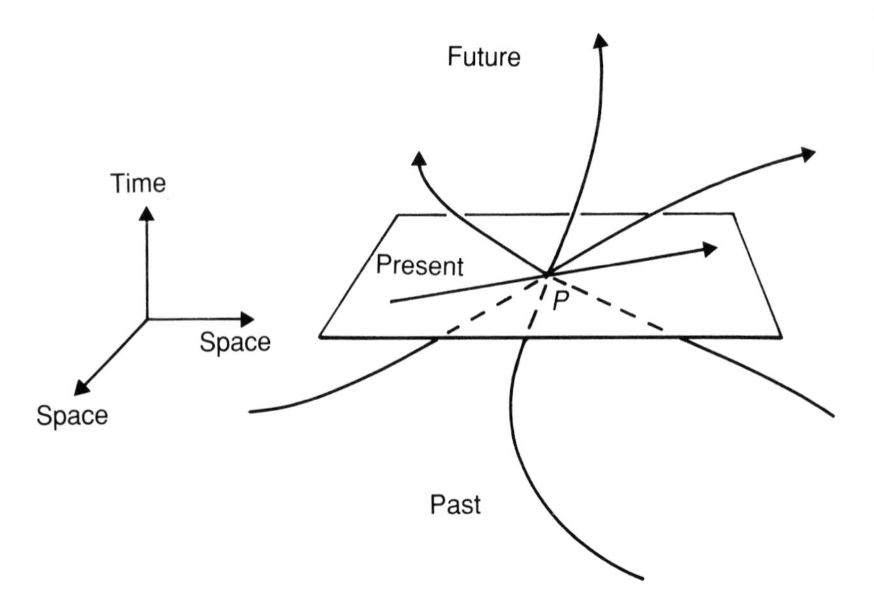

Figure 3. The causal structure of Newton's space-time continuum (one space dimension has been suppressed). Each point P is characterised by a universal time, simultaneous throughout space. The trajectories of bodies travel from the past to the future without restriction, except in the extreme case of bodies moving infinitely fast and remaining at constant time.

A vivid way of representing space-time structure which I shall use frequently is via *light cones*. Let us imagine a point in space and a light ray emitted at this point. In space empty of all matter, the front of the light wave is a perfect sphere centred on the point of emission, and this sphere increases in time at the velocity of light (Figure 4). We now suppress one of the space dimensions, so the wave can be represented by a diagram on paper. The light sphere expanding in time becomes a cone whose apex represents the place and instant (the event) when the light was first emitted. The cone itself describes the subsequent history of the light ray.

Figure 5 is a space-time diagram showing the light cones centred on several events. For a given event E, the light cone consists of two sheets, one belonging to the past and one belonging to the future. These sheets are spanned by all light rays passing through the

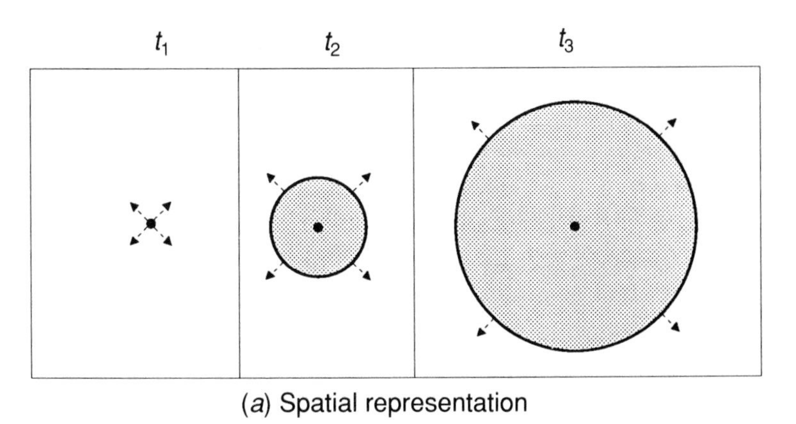

(*a*) Spatial representation

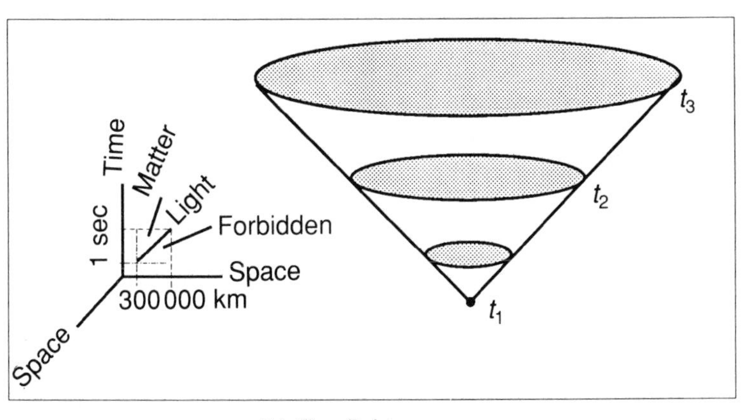

(*b*) The light cone

Figure 4. The light cone.

A flash of light is emitted at a given point. The wavefront is a sphere expanding at a velocity of 300 000 km/s, shown in (*a*) at three successive instants. The light cone (*b*) shows the entire history of the wavefront in a single space-time diagram. When a space dimension is suppressed, the sphere is represented by a circle (in projection an ellipse). The expanding circles of light generate a cone with the point of emission at its origin. In this diagram, we have chosen the units of space and time as 300 000 kilometres and 1 second respectively, so the light rays follow lines inclined at 45°.

event, including those emitted in its past as well as those originating there.

The fundamental postulate of Special Relativity requires that no particle is able to travel faster than the velocity of light, which is an absolute constant, independent of motion. This means that in one second, a particle can travel distances of no more than 300 000 kilometres, while light travels exactly that distance. On the space-time diagram, this property is shown by the fact that all particles have a *worldline* (the name given to a space-time trajectory) situated *inside* the light cone and that at the limit the worldlines of photons (particles of light) are exactly on the light cone, because they generate it.

The causal structure of Minkowski's space-time is thus very different from Newtonian space-time, because in it the velocity of light is the maximum transmission velocity of any signal. For each event *E*, the light cone divides space and time into two parts: events which can be influenced by an electromagnetic signal coming from *E* (the inside of the cone) and those which cannot (outside the cone or 'Elsewhere'). Special Relativity forbids a worldline from leaving the confines of the light cone and penetrating the Elsewhere, and vice versa.[1]

In summary, the trajectories of light rays enable us to visualise the *framework of the space-time continuum.* In Special Relativity, where there is no gravitation, all the light cones passing through all the events are 'parallel' to each other. Minkowski's space-time continuum is therefore rigid, or 'flat'. The clear Galilean–Newtonian separation of space and time is replaced by the concept of space-time.

Playtime

Einsteinian relativity adds the property of *elasticity* in time to that of causality. The time measured by a clock which an observer carries with him, called *proper time*, differs from that of

[1] However, this does not prevent worldlines from being situated completely in the Elsewhere. Hypothetical particles moving in this region at velocities greater than that of light, called 'tachyons', raise delicate problems of interpretation, although they have never been detected in the laboratory.

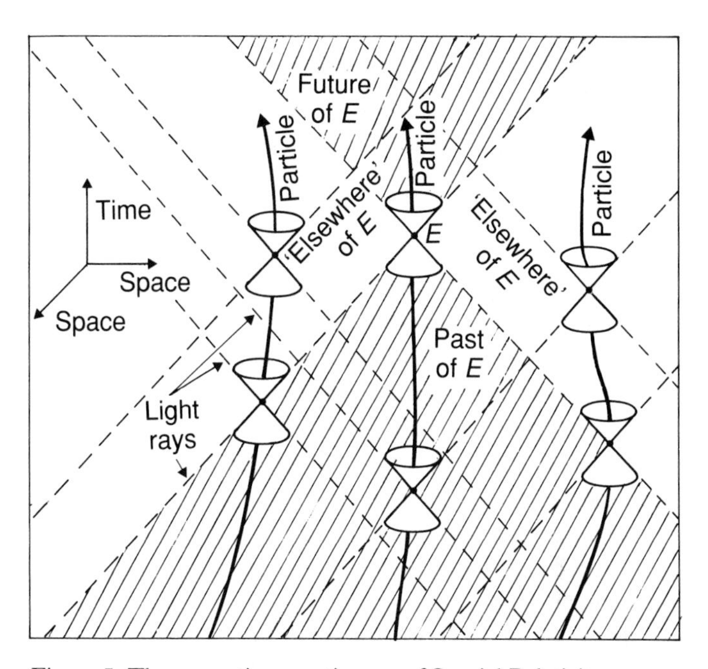

Figure 5. The space-time continuum of Special Relativity. At each event *E* in space-time light rays generate two cones. The light rays emitted by *E* span the future cone and those received by *E* span the past cone (shaded). Ordinary particles are unable to travel faster than the velocity of light, and their trajectories are confined to the interior of the cones. No light ray or particle which passes through *E* is able to penetrate the 'Elsewhere' (clear zone). The absolute invariance of the velocity of light in a vacuum is shown by the fact that all the cones have the same slope: the space-time continuum of Special Relativity, free from gravitating matter, is flat.

clocks moving relative to him. Although this is noticeable only at velocities approaching that of light, these new time rules lead to surprising situations.

Much has been written about the famous *twin paradox*. Consider a pair of twins, 20 years old; one of them undertakes a journey to explore the Universe. He makes a return journey at a constant velocity of 297 000 km/s (99% of the velocity of light), to a planet which is 20 light years away. On his return to Earth, the astronaut's watch tells him that he has been away for 6 years; however, for the twin remaining on Earth, 40 years have passed. This indeed means time as experienced by each twin is different: biological clocks are affected in the same way as atomic clocks. The brothers' ages can also be measured in terms of the number of their heartbeats: the astronaut really is only 26 when he returns and his twin is 60!

This surprising effect was explained by the French physicist Paul Langevin in 1911: of all the worldlines joining two events (in this example the spaceship's departure from and return to the Earth), that which takes the longest time is the one which is not accelerated (Figure 6). The astronaut has to accelerate and decelerate during his journey, and the situation is not symmetric; his proper time is therefore much shorter than his brother's.[2] Although it appears paradoxical, the twins' fictional experience does not indicate an internal inconsistency in Einsteinian relativity, but in fact illustrates an inescapable consequence of the elasticity of time.

Contrary to popular belief, although the theory of relativity prevents us from travelling faster than the velocity of light, it does facilitate the exploration of deep space. A variant of the twin experiment (which assumes instantaneous accelerations) assumes instead that the spaceship has a *constant* acceleration (with respect to its instantaneous inertial reference frame), equal to the acceleration of gravity at the Earth's surface and thus quite comfortable for

Figure 6 (*opposite*). Langevin's twin paradox.

The worldline with the longest proper time is the one without acceleration (vertical axis). 40 years pass for the twin who remains on Earth. His astronaut brother accelerates at the beginning of the journey, at halfway and at the end of the journey; the rest of the time he moves at constant velocity. His proper time (indicated by the numbers) is the shorter the closer he approaches the light cone (broken line: in the limit the proper time of light rays is strictly zero). The curved line shows the trajectory of the astronaut who accelerates continuously until halfway and then decelerates.

[2] The difference between the lifetimes is not solely a function of the acceleration of the traveller, but also depends on the total duration of the experiment. The accelerations are involved only in order to compare the astronaut's time with that on Earth.

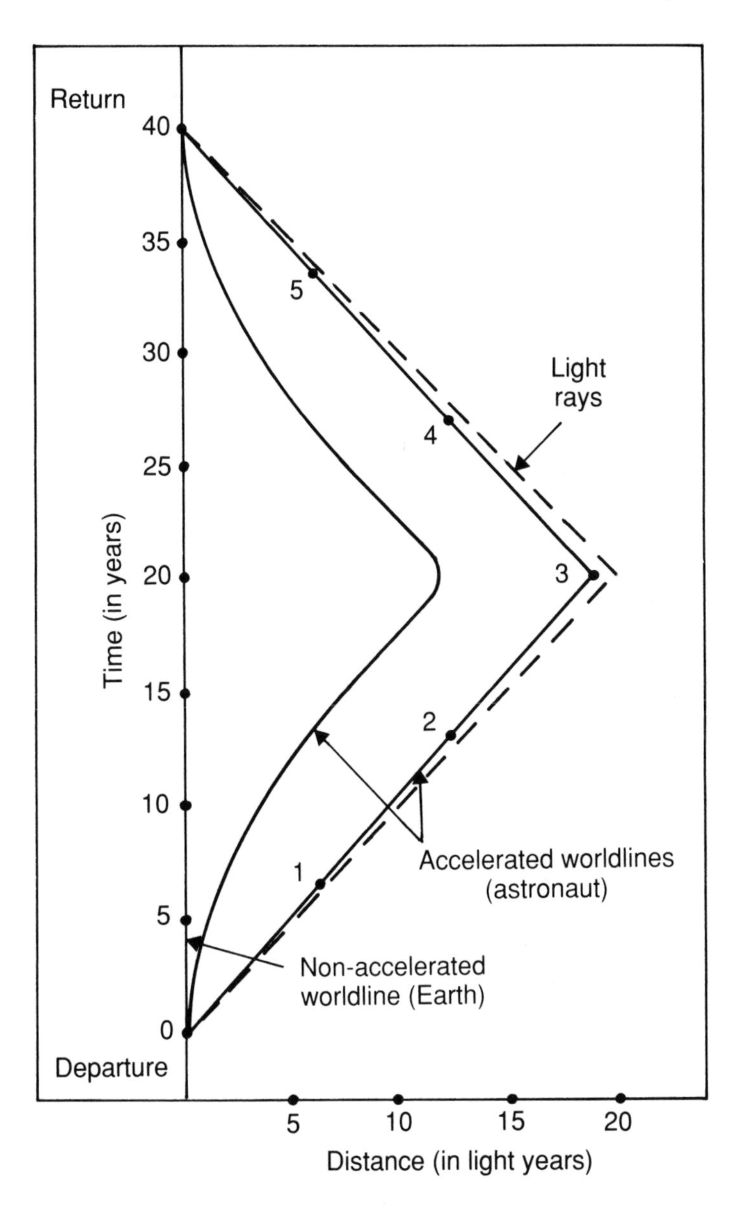
Return
40
35
30
25
20
15
10
5
0
Departure
Time (in years)
5
4
3
2
1
Light
rays
Accelerated worldlines
(astronaut)
Non-accelerated
worldline (Earth)
5
10
15
20
Distance (in light years)

the astronaut. The spaceship's velocity will rapidly increase and approach the velocity of light without ever reaching it. On board, time will pass much more slowly than on Earth. In 2.5 years measured by its own clock, the ship will reach the closest star (Alpha Centauri) which is 4 light years from Earth, and after about 4.5 years will have travelled 40 light years. The centre of the Galaxy would be reached in 10 years, but 15 000 years would have passed on Earth. In 25 years of their time (less than the lifetime of an astronaut), the ship would be able to travel once round the observable Universe, that is, *thirty thousand million light years*! It would be better therefore not to return to Earth, since the Sun would have been extinguished long before, after having burnt the planets to a cinder.

This fantastic journey is, however, impossible to realise because of the enormous amount of energy required to maintain the spaceship's acceleration. The best method would be to transform the material of the ship itself into propulsive energy. With perfectly efficient conversion, at the arrival at the centre of the Galaxy only *one billionth* of the initial mass would remain. The mountain would have shrunk to the size of a mouse.

The relativistic bomb

'If I had known, I should have become a watchmaker.'

Albert Einstein

Special Relativity is one of the best verified theories of physics. The strange phenomenon of elastic time has been demonstrated experimentally, not on human beings (a painful experiment) but using elementary particles that can be accelerated to velocities approaching that of light using reasonable amounts of energy. High precision atomic clocks have also been placed on aircraft. On their return to Earth, they are found to have measured less elapsed time than those which remained on the ground.[3] Transformation formulae between inertial reference frames, four-

[3] If someone spent 60 years of his life on board a plane flying at a velocity of 1000 km/h, he would gain only one thousandth of a second over those who remained on the ground.

dimensional space-time structure, and elasticity of time are rather abstract ideas. But Special Relativity is famous for the assertion of the equivalence of mass and energy, expressed simply as $E = mc^2$.

In 1905 the practical implications of Special Relativity, were still unguessed, but its philosophical impact was immediate. Beliefs held for thousands of years now proved ill-suited to the description of the real world. Philosophers such as Bergson refused to alter their conception of the world and viewed Einstein's theory as pure abstraction. It is a sad irony that it took the annihilation of Hiroshima by the atomic bomb to remove any remaining doubts about the validity of Special Relativity.

Special Relativity governs all phenomena involving high velocities and energies. The stream of cosmic rays which bombards the upper atmosphere creates showers of elementary particles called *mesons*, whose time of flight appears to be 50 times greater than their proper lifetimes. More importantly, Special Relativity allows us to understand why the Sun shines, converting 4 million tonnes of matter into radiative energy every second.

We see here a clear link between Special Relativity and astrophysics. However, the black holes which are the subject of this book have nothing to do with Special Relativity. They are primarily a manifestation of gravitation, while the space-time continuum of Special Relativity describes an idealised vacuum through which move only electromagnetic waves and particles whose weight is so small as to be negligible. In the real universe of stars, galaxies and black holes, all bodies are subject to gravity. To understand this, we must continue our 'demolition' of space and time. This is the challenge of *General Relativity*.

3

Curved space-time

The Principle of Equivalence

'I believe that pure thought is competent enough to comprehend the world.'

Albert Einstein (1933)

In 1905 Einstein both revived the corpuscular theory of light, and rendered Maxwell's electromagnetic theory coherent. But he found himself on the horns of a dilemma. The two aspects of radiation were contradictory: if light was composed of material corpuscles it would be influenced by matter, by virtue of the law of universal attraction between bodies; but if this was true, how could the velocity of light be the absolute constant c required by Special Relativity?

It is of course gravity which is responsible for this conflict. Gravity is ubiquitous in the Universe and *accelerates* all matter, whereas the inertial reference frames of Special Relativity are precisely those which are not accelerated. Plainly gravitation is ignored by Special Relativity. Einstein was well aware of this anomaly, and understood that before gravitational forces could be incorporated into Special Relativity's electromagnetic space-time a new understanding of the concept of 'force' itself was required.

Newton's law of universal attraction required that all bodies possessed an intrinsic characteristic called *gravitational mass*. This was a measure of the gravitational force exerted by each material body. In addition, Newton summed up in three basic laws how material bodies reacted when subjected to *any forces* whatever, gravitational or not. The first law was simply Descartes' Principle of Inertia: in the absence of a force, a body either remains at rest or moves with uniform velocity in a straight line. The second law stipulates that an accelerated body is subjected to a force proportional to its acceleration and its mass (the well-known formula, $F = ma$). The third law expresses the equality of action and reaction: each force acting on a body (for instance a man pushing against a wall) is accompanied by an equal force in the opposite direction (the wall pushing against the man).

The Newtonian force is therefore that which causes a body to deviate from its inertial motion, and the resistance of the body to all changes in its state of inertia is measured by its *inertial mass*. In this way of thinking, the force of universal attraction is a force like any other, and the gravitational mass characterising it is to gravity what an electric charge is to the electric field. We know, for example, that certain bodies with the same inertial mass but different electric charges are accelerated differently in the same electric field. Thus in Newtonian theory, there is no reason for the gravitational mass and the inertial mass to be identical.

But the fundamental property of gravity, observed by Galileo and Newton, is that terrestrial gravity accelerates all material bodies *in the same way*, independently of their inertial or gravitational mass, their size or their nature. A feather, a molecule or a tonne of bricks released near the Earth's surface all accelerate at the same rate of 9.8 m/s^2.[1]

In other words, not only are there no 'gravitationally neutral' bodies, but all material bodies carry exactly the same relative gravitational charge. This can only be possible if the gravitational mass and the inertial mass are *identical*. This property was adopted as axiomatic and called the *Principle of Equivalence*.

What was initially only an approximate equivalence has become one of the most precise measurements in the whole of science.

[1] In other words, assuming there is no air resistance, then their velocity increases by 9.8 m/s each second; at the end of one second their velocity is 9.8 m/s, and at the end of two seconds it is 19.6 m/s, and so on. This constant acceleration of 9.8 m/s^2 is just the acceleration of gravity at the Earth's surface.

The Hungarian Baron Lorand von Eötvös verified the Principle of Equivalence in 1889 and again in 1922 to an accuracy of about one part in a billion. More recently, the precision has been increased 1000 times. Since all the energies in a body contribute to the inertial mass (notably the electromagnetic energy binding the electrons and nucleus in an atom), we can conclude that all energies have weight. In particular, *light has weight*.

Einstein realised that the Principle of Equivalence provided the key to understanding gravitation, a force utterly unlike the realm of electromagnetism and whose inclusion would require a radical extension of Special Relativity. To begin, we should consider further the physical significance of the Principle of Equivalence.

For Einstein, the equivalence between gravitational and inertial mass was only a weak version of a much stronger equivalence, which united *uniform gravitation* and *acceleration* (Figure 7). Einstein actually noted that:

1. Any acceleration simulates gravity. A human being placed inside a spaceship with an acceleration equal to terrestrial gravity would notice no difference from standing on the ground.
2. The effect of gravity can be removed by choosing a suitably accelerated reference frame. His favourite example was that of a lift with its cable cut, inside which an observer would have the same sensation of *weightlessness* that he would experience in space, freed from the influence of the Earth's gravitational pull.

We see here the vast difference between gravity and all other forces of Nature, for example the electric force. It is impossible to simulate the electric force by an acceleration, since bodies placed in an electric field are not all subjected to the same acceleration (this depends on their charge as well). In other words, gravitation is not really a force exerted between different bodies in space-time, but a property of space-time itself.

This shattering intrusion of gravitation into the intimate structure of space-time is the theory of General Relativity.

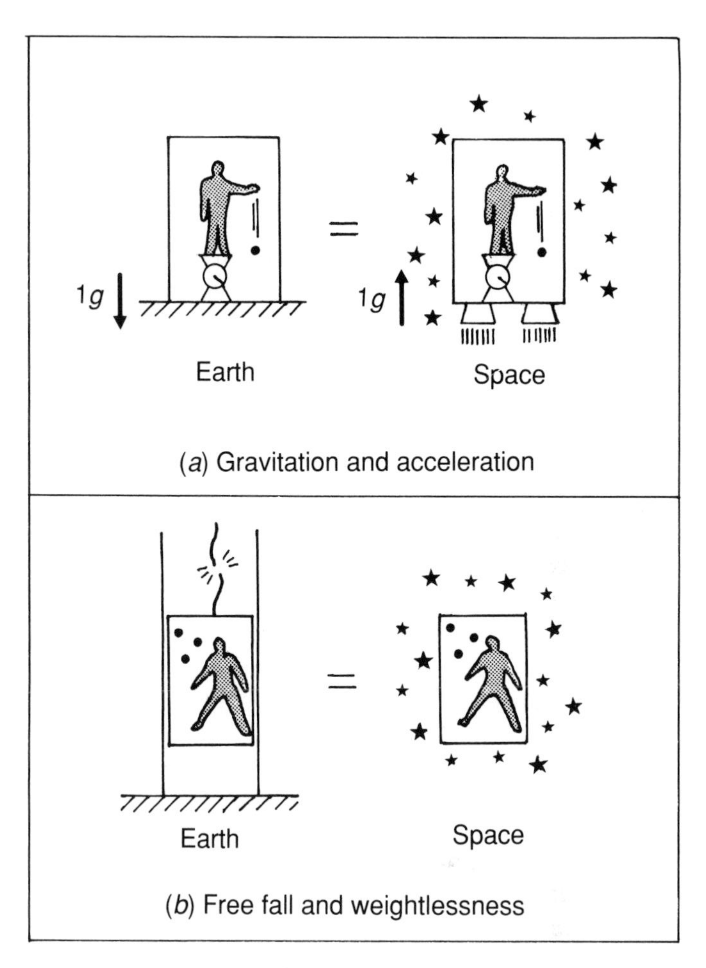

Figure 7. The Principle of Equivalence.

The new inertia

We recall that consistent physics requires a form of relativity specifying the nature of the reference frames in which physical laws take a specific form. In this sense, it is clear that General Relativity overthrows Special Relativity. In Special Relativity all reference frames move with constant velocities, free of all forces or accelerations. The space-time continuum is a flat desert, without any local characteristics, and this emptiness ensures the relativity of positions and velocities. But in the presence of gravity, all reference frames are accelerated. There is therefore no universal inertial reference frame in General Relativity. The space-time continuum deforms and acquires hollows, and positions and velocities can be specified relative to these. All reference frames, inertial or not, can be used in describing the laws of Nature, provided we know how to pass correctly from one to another. In this sense the name of Einstein's gravitational theory is badly chosen because General Relativity is less relative than Special Relativity.

Since a uniform gravitational field can be eliminated or simulated by an acceleration, and vice versa, a body falling in such a field is *free* of all forces (in the case of terrestrial gravity a man is prevented from falling towards the centre of the Earth by the ground which exerts pressure under his feet). Free fall in a constant gravitational field is therefore the 'natural' motion of a body. In any region of the Universe sufficiently small that gravity does not vary much, the motion of free fall defines a *local* inertial reference frame in which the laws of physics take their simplest form, described by Special Relativity. There is no question of abandoning Special Relativity: on the contrary, it is supplanted by a more general theory but remains applicable within a certain domain of validity.

The cosmic golf course

We know today that *space-time is curved*, but what does this strange and fascinating statement actually mean?

The twin paradox is a good illustration of how the rigid structure of space-time in Special Relativity lets space and time be subject to separate distortions (contraction or dilation), caused by the motion of observers. General Relativity takes on its full significance, revolutionising our view of the Universe, in asserting that gravity distorts the entirety of space-time.

If, at a given point, there is no direct effect of gravitation, we can measure the differential effects between neighbouring points. In a lift where the cable has been severed, two 'free' bodies follow parallel trajectories to first approximation, but in reality the two trajectories intersect at the centre of the Earth, 6400 kilometres

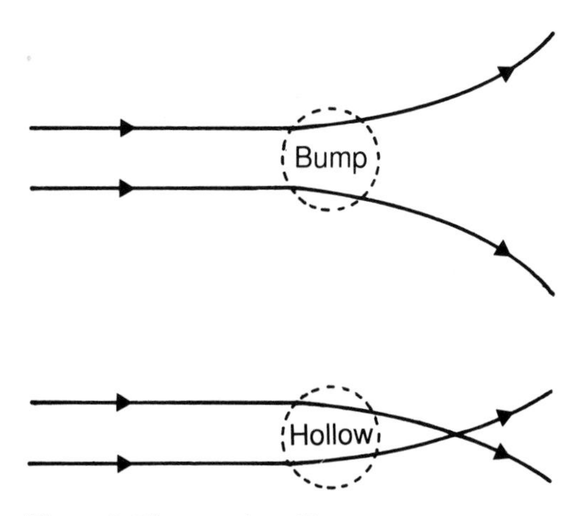

Figure 8. The cosmic golf course.

The motion of two golf balls, with initially parallel trajectories on a flat terrain. On encountering a hump they move apart and on encountering a hollow they move closer together.

away. There is therefore a relative acceleration between the two trajectories (since they are approaching each other), corresponding to a *differential* gravitational field.

A striking illustration of the difference between direct and differential gravitation is the amplitude of ocean tides. Although the direct gravitation from the Sun on the surface of the Earth is 180 times stronger than the Moon's, the solar tides are much lower than the lunar tides. This is because the tides do not result from direct gravitation but from the variations in the solar and lunar gravitational forces at different points on the globe. This variation is 6% for the Moon and only 1.7% for the Sun.

In Newtonian theory differential gravitational effects appear as *tidal forces*. In the Solar System tidal forces are very weak, but those caused by black holes are capable of destroying entire stars (see Chapter 17). However, the description of differential gravitation in terms of tidal forces is completely superfluous to General Relativity: its effects are not mechanical but purely *geometrical*. To understand this, let us consider two nearby golf balls initially following parallel trajectories (Figure 8). If the terrain is perfectly flat, the trajectories will remain parallel. If not, their relative positions will change; a hump will cause them to part and a hollow will throw them together. On a cosmic golf course, differential gravitation can be represented by a curve on the space-time 'green'. And, since gravitation is always attractive, the curves will always be hollows and not humps.

The profound significance of the curvature of space-time is thus the relationship between gravitation and geometry, imposed by the Principle of Equivalence. Material bodies are not compelled to move through a 'flat' space-time under the action of gravitational forces, but freely follow the contours of curved space-time.

Curved geometries

'God writes straight with curves.'

Freemason thinker (1782)

The word 'curvature' is part of our everyday vocabulary. Euclidean geometry in three-dimensional space allows us to speak of the curvature of lines or surfaces, which have only one or two dimensions. The circle, a geometrical figure of one dimension (it has 'length', but no 'width' or 'depth'), is more curved if its radius is shorter. Conversely, if the radius is increased to infinite length, a circle becomes a straight line, straightening itself out by losing its curvature. In the same way, a sphere becomes a plane as its radius is increased without limit (on local scales we see the Earth's surface as flat, apart from surface roughness).

Curvature has therefore a precise geometric definition, but when we increase the number of dimensions the definition becomes more complicated, and curvature cannot be described by one number as in the case of the circle; we must now speak of 'curvatures'. Let us examine the simple case of a *cylinder*, a two-dimensional surface (Figure 9). The curvature measured parallel to its axis of symmetry is simply zero, whereas the curvature measured in the direction perpendicular to it is that of the circular cross-section of the cylinder.

Despite the multiple facets of curvature, it is possible to speak of an *intrinsic* curvature. At each point on a two-dimensional surface

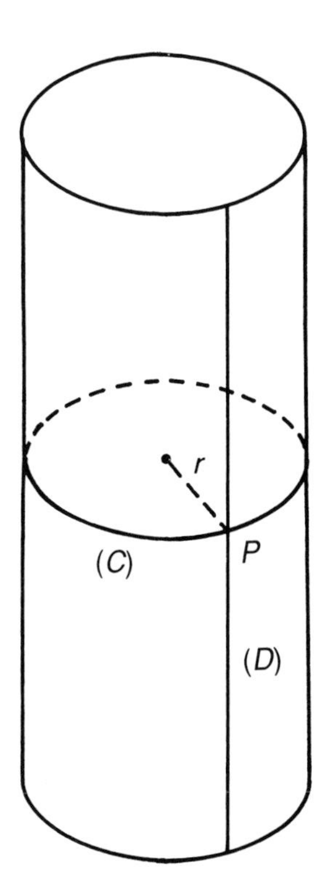

Figure 9. The curvature of a cylinder.

At a point P two numbers are required to specify the curvature. Parallel to the cylinder's axis (D), the curvature is zero; in a perpendicular direction , it is equal to that of the inscribed circle (C).

we can measure radii of curvature in two perpendicular directions whose inverse product gives the intrinsic curvature of the surface. If the two radii of curvature are on the same side of the surface the curvature is *positive*, and if the radii of curvature are on opposite sides the curvature is *negative*. The cylinder therefore has zero intrinsic curvature; in fact it can be sliced and placed flat on a table without tearing it, something which cannot be done with a sphere (Figure 10).

Visualising a sphere, cylinder and two-dimensional surfaces in general as being 'embedded' in three-dimensional Euclidean space

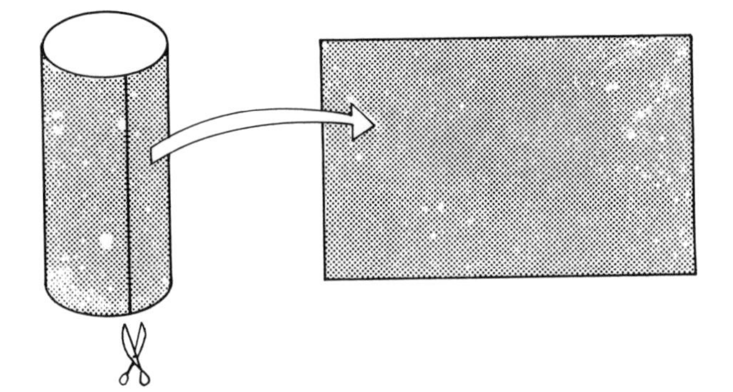

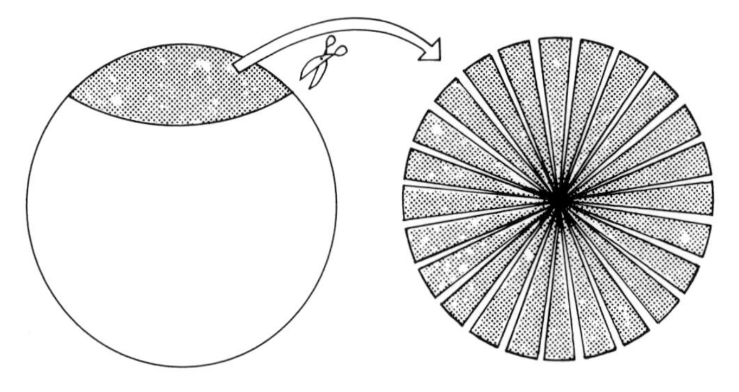

Figure 10. The cylinder and the sphere.

The intrinsic curvature of a cylinder is zero: it can be cut and placed flat on a piece of paper. The curvature of a sphere is strictly positive; if we cut out a circular cap and try to flatten it, its surface tears and its area is less than that of a circle with the same radius on a plane.

encourages us to define an 'interior' and 'exterior', and to say that a surface is curved 'in something'. However, in purely geometrical terms, all the properties of two-dimensional surfaces can be measured without any knowledge of the space which contains them. This remains true no matter how many dimensions there are. *The curved geometry of the four-dimensional Universe can be described without leaving it or referring to some hypothetical larger space.* Let us see how.

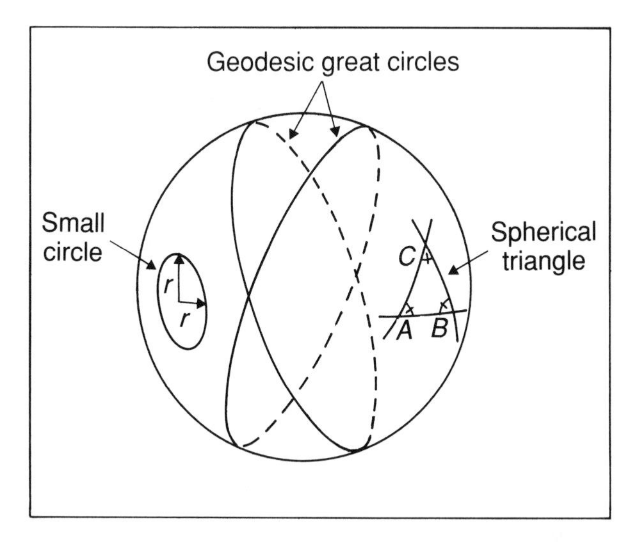

Figure 11. Spherical geometry.

The sphere is a two-dimensional closed surface with uniform positive curvature (the same at each point). The geodesics (the shortest trajectories, equivalent to lines on a plane) are great circles, centred at the origin and having the same radius as the sphere. There are no parallel geodesics therefore. The intrinsic geometry of a sphere is not Euclidean. The sum of the angles of a spherical triangle *ABC* is greater than that of the angles of a plane triangle (180°). The ratio of the circumference of a small circle to its radius r is less than 2π.

The mathematical theory of curved space was developed in the nineteenth century, principally by Bernhard Riemann. Even in the simplest cases, curved geometries have properties unknown in Euclidean geometry.

Let us consider the surface of a sphere once more (Figure 11). It consists of a two-dimensional space whose curvature is positive and uniform (the same at each point), since the two radii of curvature are equal to the radius of the sphere. The shortest route joining two distinct points on the sphere's surface is an arc of a *great circle*, i.e. a portion of a circle with the same centre as the sphere drawn on its surface. Great circles are to the sphere what straight lines are to the plane: they are *geodesics*, that is, curves of shortest length. A pilot in a hurry to fly from Paris to Tokyo without stopping (or avoiding

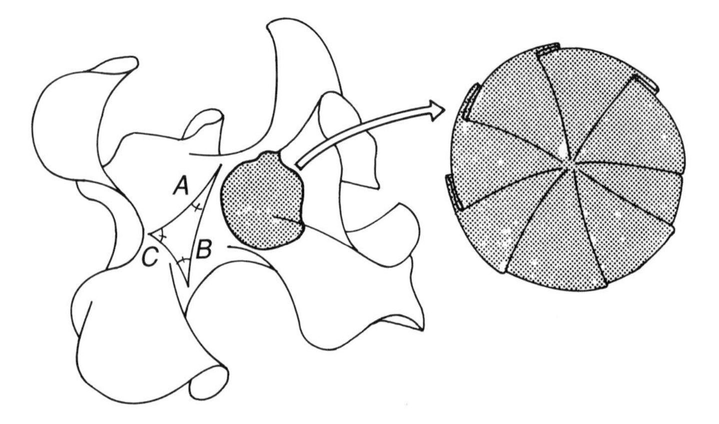

Figure 12. Hyperbolic geometry.

Its curvature is negative. The sum of the interior angles of a hyperbolic triangle *ABC* is less than 180°. If a circle is cut from the surface and laid flat on a table, it creases, and its area is greater than that of a circle of the same radius cut from a plane.

forbidden air space) would fly north, passing over Siberia and then flying south, to follow the shortest trajectory. Since all great circles are concentric, any two of them will intersect at two places (for example, the meridians meet at the poles). In other words, on the surface of the sphere there are no parallel 'straight lines'.

Here we see Euclidean geometry being thoroughly maltreated! The well-known laws of Euclidean geometry apply to 'flat' space, without any curvature. As soon as there is any curvature, these familiar laws are completely overthrown. The most obvious geometric property of the sphere is as follows: all lines in a plane extend to infinity, but if one moves in a straight line on a sphere (i.e. along a great circle), one always returns to the point of departure from the opposite direction. A spherical surface is therefore *finite*, or *closed*, although it has no limit or edge (none of the great circles have an end). The sphere is the idealised prototype of a finite space of any number of dimensions.[2]

Let us now examine the case of space with negative curvature.

[2] The Earth's surface, deformed by rotation, terrain and tides is not exactly spherical, but it has the same properties.

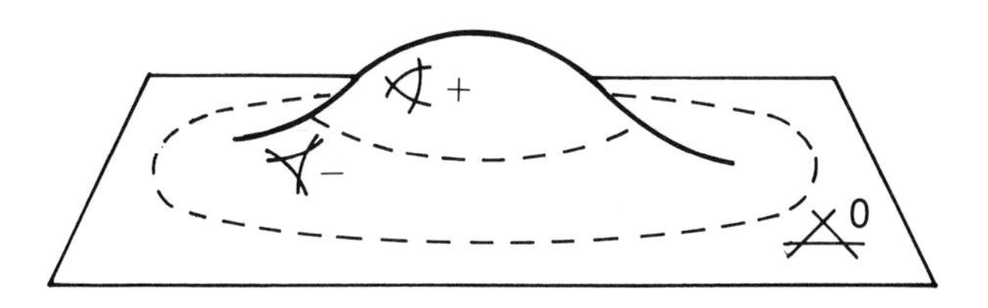

Figure 13. A surface of variable curvature.

Restricting ourselves to two dimensions for simplicity, the classic example is a *hyperboloid*, a saddle shape (Figure 12). If we move in a straight line along the surface, we in general never return to the point of departure but instead continue indefinitely to depart from it. Like a plane, a hyperboloid is an *open* surface, but the similarity ends there, for, being curved, the hyperboloid is not Euclidean at all.

In general most surfaces do not have uniformly positive or negative curvature, like the sphere or hyperboloid. The value of the curvature varies from point to point and can change its sign from one region to another (Figure 13).

Geometry and matter

'Ubi materia, ibi geometria.'

Johannes Kepler

We now consider the four-dimensional geometry of General Relativity. It is essential to understand that *space-time* is curved and not just space. Riemann had already tried to make electromagnetism and gravitation compatible by considering twisted space; he failed because he did not 'wring the neck' of time as well.

Let us suppose that we want to throw various projectiles at a target, 10 metres away on the ground. Under the Earth's gravity the projectiles all follow parabolic trajectories passing through both the thrower and the target, but their maximum heights depend on their initial velocities (Figure 14*a*). If a ball is thrown at 10 m/s, it will reach the target 1.5 seconds later by means of an arc of 3 metres height. If a bullet is fired at the target with a velocity of 500 m/s, it will take 0.02 second to reach the target, following an arc of 0.5 millimetre; if the bullet is fired 12 kilometres into the air and allowed to fall back on to the target,[3] it will reach the target after about 100 seconds. In the limit, the target can also be reached by a light ray travelling at a velocity of 300 000 km/s. In this case the arc is imperceptible, and the trajectory through space is almost a straight line. It is obvious that the radii of curvature of these parabolic arcs are very different.

Let us now add the time dimensions (Figure 14*b*). The radii of curvature as measured in space-time are all *exactly the same*, whether for the ball, the bullet or the photon: they are of the order of one light year. It is therefore more reasonable to say that the space-time trajectories are 'straight' and that space-time itself is curved by terrestrial gravitation. The projectiles which are not subject to any forces would then travel along geodesics (the equivalent of 'travelling in a straight line' in a curved geometry).

The previous example shows how space-time is much more curved in time than space. This temporal curvature becomes apparent as soon as the velocities involved start to increase. A hump on the road, a small irregularity in the spatial curvature, is hardly noticed by a pedestrian walking slowly. However, for a car travelling at 120 km/h, the hump is a danger; it causes a much greater distortion in the temporal dimension.

Arthur Eddington calculated that a mass of 1 tonne placed at the centre of a circle of radius 5 metres would modify the curvature of space only enough to alter the ratio of circumference to diameter (the number π in Euclidean geometry) in the 24th decimal place.

Thus an enormous mass would be required to alter space-time in an appreciable manner. The fact that the space-time radius of curvature on the surface of the Earth is so large (about 1 light year, i.e. a billion times the radius of the planet), implies that the Earth's gravitational field, accelerating bodies at a rate of 9.8 m/s^2, is very modest. For the vast majority of physical experiments near the Earth we may continue to use Minkowski's flat space-time and

[3] Neglecting atmospheric perturbations and the Earth's rotation!

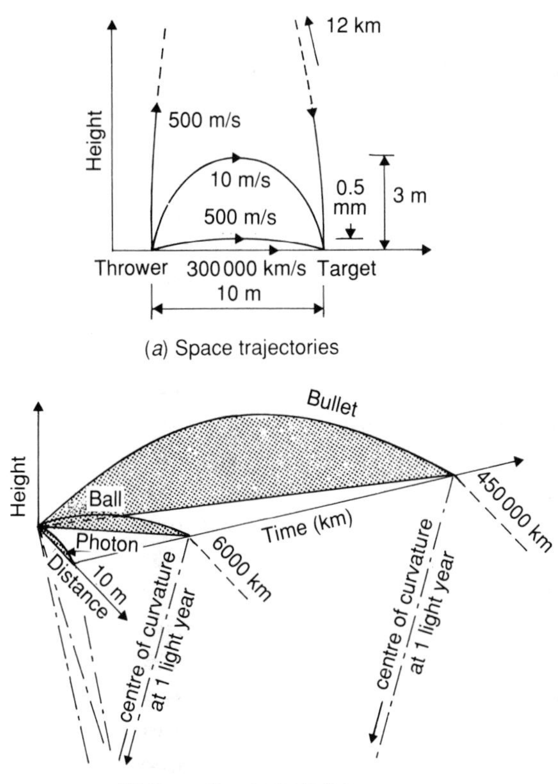

Figure 14. Space-time curvature.
Diagram (*a*) illustrates purely spatial trajectories; the curvatures are very different. Diagram (*b*) represents the space-time trajectories. Since light travels at 300 000 km/s, we can measure time in kilometres: the ball reaches the target in a time of 450 000 kilometres (1.5 seconds), the bullet reaches the target in 6000 kilometres (0.02 second). The curvatures of the trajectories of the photon, bullet and ball become identical. The curvature is caused by the Earth's gravitational field. (After Misner, Thorne and Wheeler.)

Special Relativity; Euclidean space and Newtonian mechanics are accurate enough if the velocities involved are small.

Despite appearing locally flat, the Universe actually is deformed by matter. However, the effects of the curvature only become significant in the vicinity of large concentrated masses (for example, black holes), or on a very large scale (several million light years) when considering clusters of thousands of galaxies. The recent discovery of *multiple quasars* is a perfect demonstration of the reality of curved space-time; the light rays emitted by a distant source have followed different optical paths through the curved space-time, providing astronomers with several images of a single object.

The mollusc of light

'Light . . . more light!'

Goethe's last words (1832)

The rigid structure of the space-time of Special Relativity – like that of Newtonian space – is completely destroyed by the impact of gravitation. The space-time continuum is *soft*, deformed by the matter it contains, and the matter in it moves according to its curvature.

The trajectories of the light rays nevertheless continue to follow the shortest paths. The frame of this space-time 'mollusc' is still woven by light, and its representation by *light cones* still summarises the essence of General Relativity (Figure 15).

Another helpful way of visualising curved space-time and its influence on matter uses a *rubber sheet*. Imagine a portion of space-time reduced to two dimensions, made of an elastic material. In the absence of any other object, the material remains flat. If a ball is placed on it, the material deforms, making a hollow around the ball that is the deeper the greater the mass of the ball. This type of representation, which appears fanciful, can be made mathematically rigorous by what are called *embedding diagrams*, which I will discuss in more detail to explain certain strange properties of black holes (Chapter 12). Figure 16 uses this representation to illustrate the deviation of light rays passing close to the Sun, and the consequences for the apparent position of stars during eclipses.

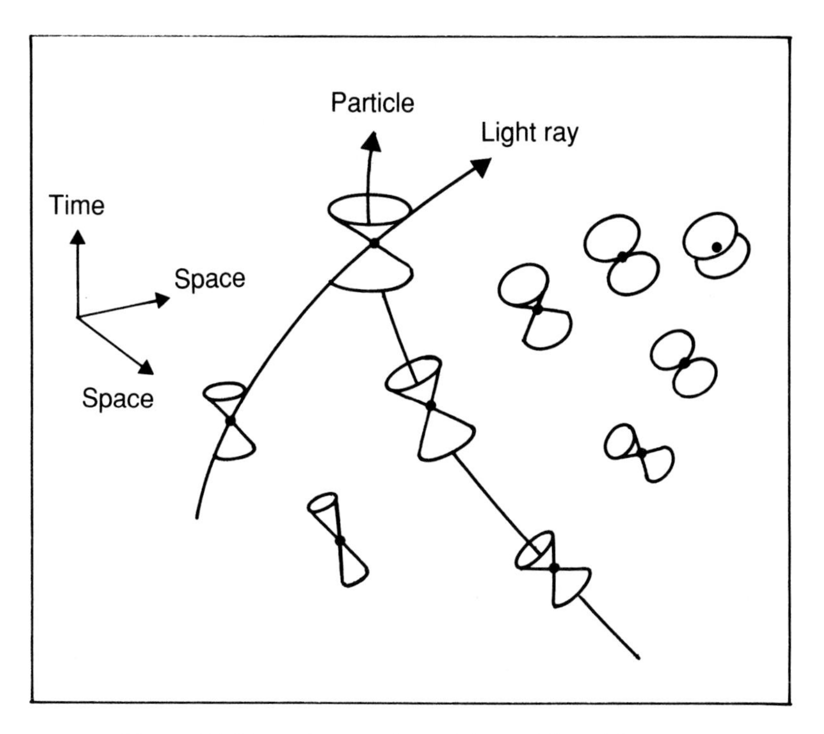

Figure 15. The soft space-time of General Relativity.

The light cones bend and deform themselves according to the degree of curvature, transmitting the Principle of Equivalence and the influence of gravity to all forms of energy. Special Relativity remains locally valid, however: the trajectories of particles through space-time are confined to the interior of light cones.

Einstein's equations

'The idea that physicists would in future have to study the theory of tensors created real panic amongst them following the first announcement that Einstein's predictions had been verified.'

A. Whitehead (1920)

All theories have their equations. Einstein's *gravitational field equations* relate the degree of distortion in space and time to the nature and motion of the sources of gravitation. Matter tells space-time how it must curve; space-time tells matter how it must move.

Einstein's equations are extremely complex. The quantities involved are no longer just forces and accelerations, but also distances and durations. They are *tensors*, each a kind of table with several entries containing all the information about geometry and matter.

The action of gravitation on matter is more complicated than that of an electric field, and requires more complex mathematical entities to describe it than numbers and vectors with three components. To be convinced of this we remember that in Newton's gravitational theory only the gravitational mass of a body is a source of gravitation, this mass being represented by a single number intrinsically associated with the body. In Einstein's theory, the gravitational mass is only one of the components of the total quantity of gravity associated with a body. Special Relativity (which is always valid in a small region of space-time, where gravitation is uniform) already shows that all forms of energy are equivalent to mass, and hence gravitating. Now the energy of a body depends on the relative motion of the observer measuring it. In a stationary body, all the energy is contained in its 'rest mass' ($E = mc^2$!); but as soon as the body moves its kinetic energy will create mass and thus gravitation. To evaluate the gravitational effect of a body, it is therefore necessary to combine its energy at rest with a 'momentum vector' describing its motion. This is why the full description of sources of gravity uses the 'energy-momentum tensor'.

Furthermore, at each point in space-time 20 numbers are required to describe the curvature. The geometric deformations of space and time therefore require the use of a 'curvature tensor' (we remember that the curvature becomes more and more complicated as the number of dimensions increases). Einstein's equations simply describe the relationship between the curvature tensor and the energy-momentum tensor, placing them on either side of an equality: matter creates curvature and curvature makes matter move.

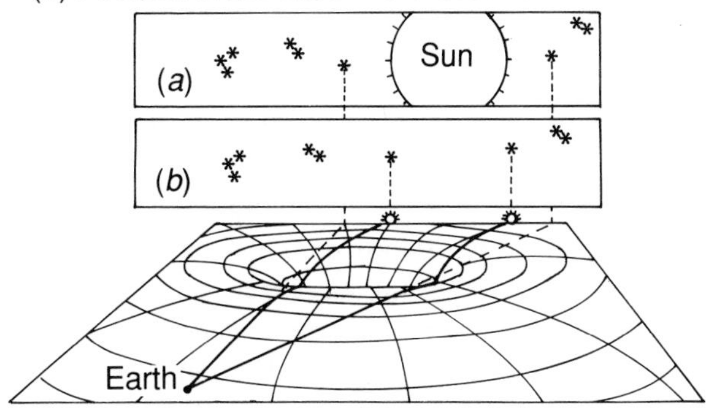

Figure 16. Light rays deviated by the Sun.

When the Sun lies between the observer and distant stars it curves the intervening space and deviates light rays, forcing them to follow the curvature. The apparent direction of the stars is shifted from their true position. (After Misner, Thorne and Wheeler.)

This is not the book to spell out all the richness of Einstein's equations. The different components of the curvature tensors and the energy-momentum tensors are so tightly interlinked that *in general it is not possible to find an exact solution*, or even to define globally what is space and what is time. Also, we have to idealise the sources of gravitation in order to calculate 'something'; to the point that most of the solutions found (describing curved space-times) bear no relation to real space and time. Einstein's equations are in some sense too prolific, and allow an infinity of theoretical universes with bizarre properties.

This richness has perhaps harmed the credibility of Einstein's theory. However, we should not get the idea that General Relativity only predicts properties which cannot be observed or are beyond human understanding. On the contrary, Einstein was both a physicist and a philosopher, and accordingly tried to describe the Universe, beginning with the Solar System. Using approximate solutions of his equations, he first calculated three measurable effects of gravitation in the Solar System not predicted by Newton's law of attraction: the deviation of light rays passing close to the Sun, anomalies in the orbit of Mercury, and the lowering of electromagnetic frequencies in a gravitational field. The next section will discuss the success of these three predictions of General Relativity.

Besides these cases, there are naturally-occurring situations where the simplifications imposed on the sources of gravitation are perfectly justified; the resulting exact solutions to Einstein's equations give a satisfactory description of some part or other of the Universe. Paradoxically, these simplifications are most fruitful at two extreme distance scales. We can calculate the gravitational field produced by an isolated body in a vacuum (i.e. the space-time distortions around the body). The region around a star – for example the Solar System – or that near a black hole, is well described by this solution, as the matter is effectively concentrated in a small region of space-time surrounded by near-vacuum. At the opposite extreme, we can calculate the average gravitational field of the Universe as a whole (its geometry), because on a very large scale matter is spread more or less evenly and galaxies act like molecules in a homogeneous cosmic gas. General Relativity thus allows us to do *cosmology*, that is, to study the shape and evolution of the Universe in its entirety. Moreover until the advent of relativistic astrophysics in the 1970s, cosmology was the only real field of application for General Relativity – along with black holes, of course.

The third major application of General Relativity, *gravitational waves*, will probably have to wait until the twenty-first century. Einstein's equations play a role in gravitation which is analogous to the role that Maxwell's equations play in electromagnetism. Now it is known that the acceleration of electric charges generates electromagnetic waves. Similarly, General Relativity also predicts that the motion of gravitational sources will cause waves, ripples of curvature moving through the elastic fabric of space and time at the velocity of light. I will discuss gravitational waves in more detail in Chapter 18.

Testing General Relativity

'In many aspects, the theoretical physicist is merely a philosopher in a working suit.'

P. Bergmann (1949)

The three observations proposed by Einstein as checks of General Relativity were the deviation of light rays near the Sun, anomalies in Mercury's orbit and the redshift of atomic spectral lines in a gravitational field.

The deviation of light passing near to the Sun (illustrated in Figure 16) was measured during the solar eclipse of 1919, and these measurements agreed with the value that Einstein had calculated.

The second test concerns planetary motion. Newton's celestial mechanics states that an isolated planet moving round the Sun will have a fixed elliptical orbit (with a major axis that does not move). In the presence of other planets this motion is perturbed, and the elliptical orbit advances slowly. In 1859, the French astronomer Urbain Le Verrier discovered that the *perihelion* of Mercury (the point of the orbit closest to the Sun) advanced more rapidly than Newton's theory predicted (Figure 17). Detailed calculations of the perturbations caused by the outer planets (essentially Jupiter) predict an advance of 5514 arcseconds per century, while Mercury actually advances 5557 arcseconds per century, 43 arcseconds too much.[4] Obviously this anomaly is quite small (at this rate it would take Mercury 3 million years before it was a whole revolution ahead), but Newton's theory is so precise in its domain of applicability that it should have provided an explanation.

The most natural assumption is to invoke the existence of a perturbing body: a ring of matter in orbit around the Sun, or even an unknown planet. This type of consideration had already made Le Verrier famous, since the analysis of perturbations in Uranus' orbit had in 1846 enabled him to predict the existence of another planet, Neptune, discovered shortly afterwards. Seeking to repeat

[4] A circle is an arc of 360° and each degree contains 3600 arcseconds.

his exploit, Le Verrier predicted the existence of another planet between the Sun and Mercury, which he called Vulcan. Le Verrier calculated that Vulcan would pass only infrequently across the Sun's disc (the dark projected spot giving the only hope of detection) but died in 1877 just before the predicted transit. He never knew of his failure; on the appointed day, every telescope was trained on the Sun, but Vulcan obstinately refused to appear.

A number of slightly modified Newtonian theories of gravitation were developed, with the sole aim of explaining the advance of the perihelion of Mercury. It was already known that other planets showed anomalous perihelion advances, for example Venus, Earth, and the asteroid Icarus, but theories which could explain the phenomenon for one planet did not work for the others.

Eventually, realising that the planets showing perihelion advances were those closest to the Sun, astronomers began looking for a perturbing force originating within the Sun itself. Clearly, the Sun is not exactly spherical, and a deformation can, in principle, cause a perihelion advance. In reality, the Sun is too round. Newtonian theories of gravitation, modified or not, remained stubbornly frustrated by the whims of a handful of planets.

In 1916 Einstein's General Relativity finally offered a coherent and unified explanation of the perihelion advances of the planets. They are not drawn by a mysterious attractive force emanating from the Sun, but move *freely* in space-time curved by the Sun's mass. Their trajectories are geodesics; the geodesics of a space-time continuum curved by a solar mass are not exactly ellipses or hyperbolae; their axes advance slowly in time by an amount precisely equal to that observed (Figure 18).

The third test proposed by Einstein concerns the apparent 'slowing down' of light in a gravitational field. The reduction in the frequency of electromagnetic radiation causes an increase in its wavelength, a 'reddening' of its spectrum (the colour red has the longest wavelength in the visible spectrum). In the Sun this effect is too weak for current experimental precision to test General Relativity. Even for stars much denser than the Sun, and thus exerting stronger restraining forces on light rays – such as white dwarfs, see Chapter 5 – the spectra are so affected by magnetic

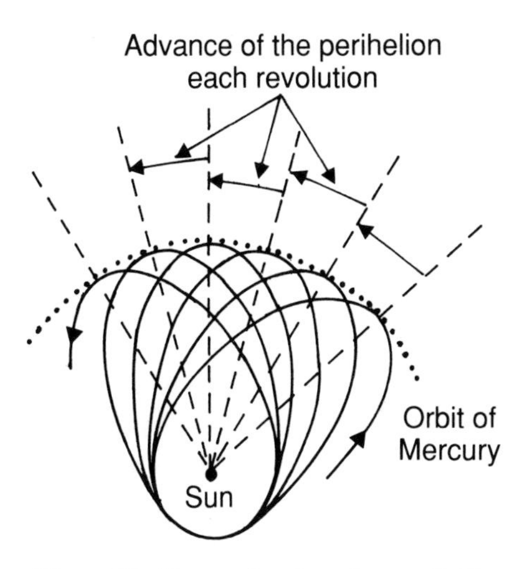

Figure 17. Mercury's advancing perihelion.

fields or the unknown motion of matter inside the star that it is difficult to correct for all the extraneous effects.

This third test is simply a version of the elasticity of time in a gravitational field. Special Relativity already shows that acceleration slows a clock down (the twin paradox). By the Principle of Equivalence, we may conclude that gravitation also slows clocks: those placed in a basement tick more slowly than those on the first floor.

It was not until after Einstein's death that it became possible to construct sufficiently accurate clocks to be able to measure the elasticity of time in a gravitational field as weak as the Earth's. In 1960, physicists at Harvard University detected with a precision of one in a thousand the frequency shift of gamma rays (high energy electromagnetic radiation) travelling down a vertical height of 23 metres. While observing the deviation of light rays passing close to the Sun requires us to wait for a solar eclipse, and checking if the perihelion of Mercury advances too fast needs a century of accumulated observations of its motion, we have here a laboratory measurement which can be repeated as desired. A flourishing era began for experimental gravitation.

From 1976, extraordinarily stable clocks, accurate to a *millionth of a billionth* part were placed on board high altitude aeroplanes, where gravity is measurably weaker than on the ground. The electromagnetic ticking of the flying clocks was compared with that of identical clocks in the laboratory. The difference between the rates was measurable and agreed perfectly with the results predicted by General Relativity.

The advent of space probes has allowed the rather larger effect of the *solar* gravitational field on the elasticity of time to be measured. A radar transmitter transmits a radio wave towards a space probe situated on the other side of the Sun; the radio wave is reflected and returned to Earth, where its journey time is measured. The geometry curved by the Sun's gravity produces a difference from the propagation time through empty flat space. The experiment was performed in 1971 with a Mariner probe and confirmed once more the occurrence of gravitational retardation.

The reader may here ask the point of accumulating expensive tests to verify a theory which seems to work so well. The reason is that in all General Relativity experiments only the gravitational field in the Solar System is involved, which is everywhere very small and stationary (it does not vary with time). Now the era of flourishing experimental gravitation excited the imagination of theorists and many theories of gravitation emerged to compete with Einstein's. Most of them contained supplementary parameters, which could be adjusted at the discretion of their inventor. This was true of the most famous of these theories, devised by the German physicist Pascual Jordan and the French physicist Yves Thiry, later revived by the Americans Carl Brans and Robert Dicke (the latter played a prominent role in the development of experimental gravitation).

Because of the freedom introduced by their parameters, the alternative theories could be adjusted in such a way as to account for all the effects measured in the Solar System. How could one then decide which was 'correct'?

Only analysis of their predictions in strong and dynamic gravitational fields (i.e. those varying rapidly in time) could provide the answer. Until very recently, nature did not provide us with a suitable test site. The discovery in 1974 of a *binary pulsar* (see Chapter 7) changed all that. The observed decay of the orbital

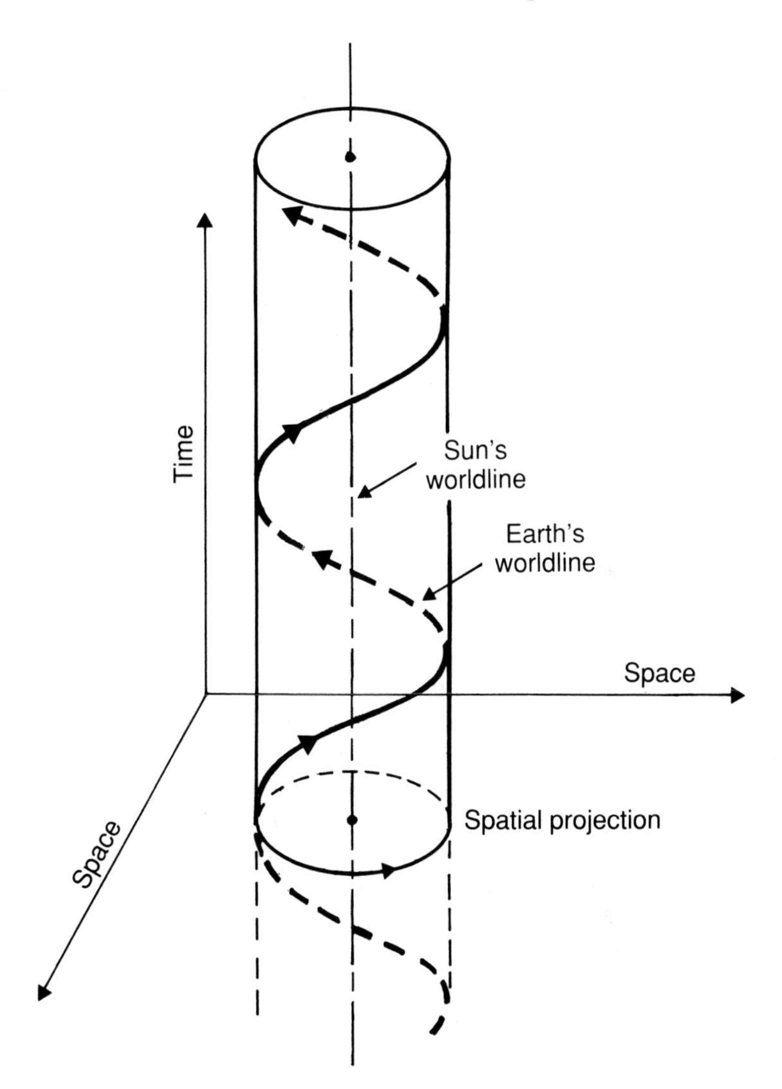

Figure 18. The motion of the Earth in space-time.

The Earth's worldline representing its motion about the Sun is a spiral wrapped around a cylinder; its spatial projection is an ellipse whose major axis slowly revolves.

period of these two closely-bound neutron stars agreed with Einstein's theory and disagreed with practically all other competing theories.

A magical theory

'The magic of this theory is such that almost no-one can escape it once he has understood it properly.'

Albert Einstein

General Relativity is certainly one of the most spectacular intellectual feats of all time, and accomplished by a single individual. In 1911, while he was working at the University of Prague, Einstein calculated for the first time the deviation of light in a gravitational field. His results were to have been verified during the 1914 eclipse, but war was declared and the project abandoned. This was fortunate for Einstein, as his theory was not quite mature and his prediction would have been in error. However, a setback would doubtless not have discouraged him. By his own admission he was a scientific 'monomaniac'. The English physicist, Paul Dirac, said later of him that 'scientific ideas dominated all Einstein's thoughts. He would offer you tea and as you were stirring it be looking for a scientific explanation for the motion of the tea leaves in the cup.'

Einstein perfected his General Relativity equations in November 1915 and published the results in the *Berliner Berichte* in the numbers dated 4, 11, 18 and 25 November. From then on his theory had a dazzling career. The first two books devoted to it appeared in 1918, one in London by Arthur Eddington,[5] the other in Berlin by Hermann Weyl. The deviation of light rays passing close to the Sun was measured during the solar eclipse on 29 May 1919 at Sobral (Brazil), thanks the zeal of Frank Dyson and Eddington. Einstein's predictions were confirmed during a

[5] At that time German science was in disfavour, and English libraries no longer took German periodicals. Eddington learnt of Einstein's articles through a Dutch friend who sent him copies by post. They were probably the only copies available in Britain.

famous meeting of the Royal Society in London on 6 November 1919.

The First World War had just finished. The world was tired and disillusioned, and looking for a new ideal. Einstein's theory with its bizarre ideas about curved space caught the imagination, although few people could understand a single word of it. Innumerable popular articles appeared in both general and philosophical journals; the public were enthralled and relativity became a fashionable conversation piece. Einstein became the most celebrated thinker in the world and his opinion was asked on all manner of subjects. The USA welcomed him with much pomp and ceremony and he became a popular figure.

In the scientific community, the enthusiasm was much more mixed. Some thinkers were overcome with admiration for Einstein's solitary creation and tried to outdo earlier eulogies of Newton. 'One of the most beautiful examples of the power of speculative thought', claimed Hermann Weyl, and did not hesitate to add, 'as though a wall obscuring the truth has collapsed'. 'The greatest exploit of human intellect', said Max Born in 1955. It is worth emphasising that among physicists the most ardent supporters of General Relativity were those able to understand it.

On the other hand those for whom the theory remained impenetrable were unrestrained. It is hard to pass over this staggering comment from the physicist H. Bouasse: 'The reason for this glory, which I believe to be ephemeral, is that Einstein's theory does not fall into the category of physical theories: it is a metaphysical hypothesis which, on top of everything, is incomprehensible, giving a double reason for its success [. . .]. Ultimately, we, the laboratory physicists, will have the last word: we accept theories which suit us; we reject those which we cannot understand and which are therefore useless to us.'

Another fierce enemy of General Relativity, Allvar Gullstrand, the Swedish opthalmologist and mathematician, winner of the Nobel prize for Physiology in 1911, was a member of the Nobel Committee for Physics. This is probably why the 1921 prize given to Einstein was 'particularly for his discovery of the law governing the photo-electric effect', and not for his theory of relativity.

The French physicist, Jean Eisenstaedt, remarked that, 'This is the fanaticism which animates honest gentlemen and makes them hate cubist, non-figurative, and dadaist paintings from the turn of the century. Those honest gentlemen who congratulate themselves on not understanding the new art that snobs applaud without understanding.'

The comparison here between scientific and artistic creativity is apt. General Relativity has often been compared to a magnificent work of abstract art. But the aesthetic beauty of a theory does not guarantee its validity, and pragmatic physicists need time to assimilate it into their canon. The International Astronomical Union (which holds meetings of astronomers from all over the world every 3 years) enthusiastically set up a 'relativistic' commission in 1922; it met once – and then decided that it was useless to carry on with its work.

Even today the battle is not over. But it has been waged stoutly, especially over almost the last 30 years, since the flickering signals of strange and distant stars first found their way into large radio telescopes.

PART 2
EXQUISITE CORPSES

'Stars are the golden fruit of a tree beyond reach.'

George Eliot

Introduction

'Science replaces the complicated visible with the simple invisible.'

Jean Perrin

Several years ago an astrophysicist opened his lecture by declaring, 'A star is something very simple.' One of his listeners immediately retorted, 'Even you would appear very simple at a distance of 100 light years!'

This remark is profoundly correct. Even though we can only investigate the Sun's 'skin', a fantastic range of phenomena can be observed: granules, sunspots, eruptions and prominences. It is only the enormous distances of other stars which reduce them to simple twinkling lights in the night sky. Only their radiation reaches the Earth, a distant and feeble echo of their prodigious internal activity. Analysis of this already gives miraculous insights, but eventually we need theoreticians of the stars to try to understand how all of it works. 'Theory' also implies 'simplification', removing inessentials so as to concentrate only on the core of the problem. I shall adopt this approach in this introduction to the abounding Universe of the stars.

In this simple view a star can be characterised in a few words: an enormous ball of hot gas. However, each of these words is significant and requires discussion.

The term 'ball of gas' implies an equilibrium. We know for

example that the Sun has not really changed for 5 billion years. This may seem surprising, since on the Earth, we are accustomed to the tendency of a free mass of gas to disperse and occupy the surrounding space. In contrast, the gas of a star, far from dispersing, remains confined in a well-determined volume. The adjective 'enormous' provides the key to this first puzzle: for a mass as large as a star, gravity dictates the organisation of the matter completely. Each atom of the star is attracted towards the centre and the mutual attraction between all the atoms ensures the cohesion of the gas; similarly gravity determines the star's shape as an almost perfect sphere provided that rotation is not too fast.

Here one might be surprised again: if all the particles belonging to a star are attracted to the centre, why does the star not collapse in on itself? The reason for this is the word 'hot': heat, that is energy, is produced at the centre of a shining star. This energy propagates towards the surface and is able to support the star's weight. On arriving at the surface, it escapes from the star as radiation.

In any discussion of stars, one word keeps recurring: gravitation, which is present at their birth but is also the cause of their death. A star's life is a permanent desperate struggle against its own weight. Permanent, because at each stage in its evolution, a star finds new resources for maintaining itself. Desperate, because the battle is already lost: sooner or later, gravity will triumph and the star will collapse in on itself.

This absolute power which gravitation exerts over the destiny of stars is repeated on a much vaster scale, shaping all the large structures in the Universe. Stars, star clusters and galaxies are born in gravitational collapse; and in gravitational collapse all of them die.

A black hole is just one of the possible stellar corpses. In my opinion, it is also the most exquisite, as it is the extreme culmination – pushed almost to absurdity – of gravitational collapse. For this reason I postpone discussion of black holes in favour of a brief résumé of the destiny of stars; how they are born, shine and die.

4

Chronicle of the twilight years

The birth of stars

Like rain, a star is a condensed droplet inside a cloud of gas. However, when conditions in space are compared to those on Earth, it can almost be said that a star forms itself out of nothing: the air we breath contains 30 billion billion atoms per cubic centimetre, while an interstellar cloud contains barely a few tens. On the other hand it spreads out over hundreds of light years and contains enough matter to form several thousand suns. The interstellar cloud also has a different chemical composition from an atmospheric cloud: it has on average 16 hydrogen atoms[1] to each helium atom and only a trace of the more complex atoms such as carbon, nitrogen and iron.

The interstellar cloud is not only rarefied, but cold: 100 K at the most.[2] Such a cloud will remain stable indefinitely, to the extent that the motions of its atoms, defining the average temperature, balance the gravitational force which tends to draw them together. Thus it is only when the cloud is *perturbed* that the star droplets can condense.

Several mechanisms can compress a cloud and trigger off the birth of stars. In the so-called *spiral* galaxies, the stars are gathered together in gigantic arms emanating from the central bulge. The

[1] Usually grouped in molecules.

[2] Degrees Kelvin (K) measure the temperature with respect to absolute zero, which is the lowest theoretically obtainable temperature. Absolute zero (0 K) is equal to −273 °C. 100 K is therefore equal to −173 °C.

arms revolve slowly around the bulge; the Sun, in the Orion arm, performs a complete revolution about the centre of the Galaxy every 200 million years. As these rotating arms transport matter, they propagate a density excess whose motion through the interstellar medium causes *compression* and triggers star formation.

Another model of star formation is based on the elegant idea that the birth or death of one star may activate the condensation of myriad new stars. When a star is born at the centre of a cloud, its intense radiation heats and compresses the periphery of the cloud causing an 'epidemic' of condensations. The cataclysmic death of a large star in a supernova[3] has a similar effect: the debris from the star, propelled at velocities of tens of thousands of kilometres per second, crashes into anything in its way and can transform the interstellar clouds into pools of young stars.

The interstellar cloud becomes opaque as soon as it starts to condense. At this point it ceases to absorb light from other stars and cools down almost to absolute zero. The atoms in the cloud are now so slowed down, almost frozen, that their mutual gravitational attraction overcomes their internal thermal motion. But the distribution of matter within the cloud is not perfectly homogeneous; there are always lumps where there are a few more atoms than elsewhere, and holes where there are a few less. Since matter causes gravitation, there is a stronger gravitational field around each of the lumps. This gravitational imbalance pulls in neighbouring atoms, which are slow because they are cold, and the attractive power increases as more and more atoms are captured. The lumps are transformed into more condensed *globules* measuring several billion kilometres and containing several stellar masses.

At this stage in the process the key mechanism, the *Jeans instability*, becomes important; in a dispersed medium, a local peak in density becomes unstable above a certain critical mass. The perturbation then separates itself from the medium to form a stable system, held together by its own gravitation. This is what happens to the globule: it is too cold to support its own weight, so it contracts and isolates itself from the rest of the cloud. As it contracts, it compresses the gas at its centre to ever increasing

[3] See Chapter 6.

pressures, temperatures and densities. The heated gas begins to radiate energy; from being black, the globule starts to glow red.

A 'star' is born, but it cannot properly be called a star because it is not radiating enough energy to support itself. The protostar therefore continues to contract, although at a much lower rate. It is only when the core temperature reaches 10 million K that the hydrogen starts to burn via *thermonuclear reactions*. This new energy pervades the core of the protostar, which stabilises: it is now a star.

The war of fire

'O Sun, it is the time of fiery reason.'

Guillaume Apollinaire

In the constant struggle against gravity, a star's main weapon is nuclear power. Its core is a nuclear bomb, constantly trying to make the star explode. It is only because the nuclear power adjusts itself to compensate gravity almost exactly that the star is able to stabilise itself for a long period of calm which may last billions of years.

As the name indicates, thermonuclear reactions occur between the nuclei of atoms at very high temperatures and thus involve the basic structure of matter. At the centre of a star like the Sun, the temperature reaches 15 million K and the pressure is 300 billion times the Earth's atmospheric pressure.[4] Under these conditions, the atoms are not only stripped of their electrons, leaving just the nuclei, but are forced to move at such velocities that they can overcome the electrical repulsion and join up, fuse. Let us examine why this *fusion* takes place.

A star is created at the centre of a cloud of molecular hydrogen and so consists mainly of hydrogen. Hydrogen is the simplest of all the chemical elements, consisting of a nucleus with positive electric charge, the *proton*, and a single *electron* carrying a negative charge, in orbit about the nucleus. In a star, the temperature is such that all the electrons have separated from the protons, which move in all

[4] Which is already 1 kilogram per square centimetre.

directions like the molecules in a gas. Since electrical charges of the same sign repel each other, the proton is 'protected' by a sort of electrical armour and keeps its distance from other protons. However, in the core of a young star, at 15 million K, the protons are moving so fast that when they bump into one another they break their armour and stick together, instead of rebounding like rubber balls.

When four protons fuse they form a *helium* nucleus. Helium is the second most abundant element in the Universe.[5] The helium nucleus weighs less than the sum of the four protons from which it was formed. This difference in mass is only a tiny fraction of the total mass (0.007), but by virtue of the mass-energy equivalence discovered by Einstein, this tiny mass loss is transformed into a colossal amount of energy. The energy liberated by converting one kilogram of hydrogen into helium is the same as that produced by burning 200 tonnes of carbon, and sufficient to keep a 100 watt light bulb lit for a million years. Stars like the Sun have enormous cores. They convert, not a kilogram, but 600 million tonnes of hydrogen into helium every second. The amount of nuclear energy is so great that its rush towards the star's exterior can hold back the gravitational contraction.

There are several possible reaction chains by which hydrogen can be transformed into helium. The most frequent reactions are the proton–proton chain (requiring only hydrogen nuclei) and the C–N–O cycle (a closed chain using heavy elements such as carbon, nitrogen and oxygen as catalysts). In the Sun, most of the energy is produced by proton–proton reactions, but in more mature stars, where the core is hotter, the reverse is true: the C–N–O cycle functions better at higher temperatures. Nevertheless, even when hot, hydrogen burns badly: in a proton–proton reaction, a proton waits on average 14 billion years before it fuses with three other protons to make a helium nucleus.[6] This 'astronomical' time

[5] On Earth, helium has practically disappeared; it is just one of the rare atmospheric gases and is used to inflate hot-air balloons. The helium created in stars is not the reason for its great abundance in the Universe. Along with hydrogen and several other light elements, most of it was formed during the first few minutes of the Universe.

[6] It takes 'only' 13 million years on average for a C–N–O reaction to occur.

explains why stars spend so long in their nuclear combustion phase and gives an indication of the enormous number of hydrogen nuclei contained in the core.

On 16 July 1945, at Alamogordo, in New Mexico, man exploded an atomic bomb for the first time. This was not really a fragment of a star as it was a *fission* bomb, in which nuclear energy is liberated by splitting much heavier nuclei than the proton. Later, man imitated the stars more closely with the hydrogen bomb, in which protons fuse together. However, there the comparison with the stars ends. The detail of the nuclear reactions is different. In a bomb, it is not necessary to wait 10 billion years for the protons to fuse; the components which are necessary for the chain reactions to work are supplied from outside, whereas in a star they are created in the core at a very low rate.

But most important of all, man is unable to control hydrogen fusion and to use it for peaceful purposes. We do not yet know how to construct a container capable of surviving the enormous pressures and temperatures involved in the reactions. The stars naturally create the furnaces which we cannot; their masses are so great that gravity confines the protons within an appropriate volume; the resulting giant nuclear reactor is stable and the energy production controlled.

The life ahead

'Bright star, would I were steadfast as thou art.'

John Keats

The energy from the Sun released at its centre is radiated as photons (light particles). But a photon has a long way to travel before it reaches the star's surface and escapes into interplanetary space, where it will ruffle comet's tails and heat the glacial crusts of planets. Contrary to what one might expect, a photon emitted at the centre of the Sun and travelling at a velocity of nearly 300 000 km/s does not take 2.3 seconds to travel the 700 000 kilometres to the surface. On average it takes 10 million years to cover this distance. The light that we receive on Earth now left the Sun's surface 8 minutes ago, but it was produced in the Sun's core when primates and mastodons walked on an Africa still separated from Eurasia.

The explanation is simple: instead of travelling in a straight line, the photon is constantly deviated from its trajectory by collisions with innumerable electrons, which are the main components of stellar matter along with protons. If the Sun's core were suddenly extinguished the light from it would continue to reach us for another 10 million years.

The stars therefore lead lives of perfectly regulated routine. Nearly all the stars seen in the sky, with the naked eye or telescope, are mature stars like the Sun, vigorously burning the hydrogen in their cores. This phase of great stability lasts 99% of a star's nuclear lifetime and is called the *Main Sequence* (see Appendix 1). Our Sun has been quietly following this Sequence for the past 5 billion years, converting its hydrogen into helium. It is halfway through its life.

Red psalm

But the Sun's 'constant' evolution will end. All burning ends in embers and extinction. When all the hydrogen is transformed into helium, the central fire loses its fuel and the quiet life of a star in the Main Sequence comes to an end. A period of major upheaval begins.

Once all the fuel has been used up there is an immediate decrease in the rate of thermonuclear reactions. The equilibrium between gravitation and the pressure of radiation is now disturbed in favour of the former. The star, with its helium core and hydrogen envelope, collapses under its own weight. The pressure, density and temperature increase; hydrogen which had remained untouched in the outer layers starts to burn, and the envelope begins to expand (unlike the core, which contracts).

Through the skilful alchemy of Nature, many elements can be transmuted into others via thermonuclear reactions. However, the heavier nuclei tend to repel each other more strongly than protons because they carry more positive charge.[7] Consequently the heavier nuclei require very high velocities to overcome their electrical repulsion and fuse together. In other words, their transmutation requires a temperature higher than 15 million K.

In this way helium nuclei in the contracting core of a star can fuse together in threes to form carbon nuclei, at temperatures of 100 million K. These in turn capture other helium nuclei and form oxygen nuclei. These new reactions have speeds utterly different from the slow destruction of hydrogen. They begin explosively,[8] and the star has to try to adjust its structure to this if possible. It takes a million years to do this. Then the flow of nuclear energy stabilises. For the next few hundred million years the star enjoys a temporary respite. Helium is consumed at its centre and hydrogen in layers further out. However, this adjustment exacts a certain price. More than the frog in the fable, the star has to expand to an enormous size to adapt its structure to the increase in its luminosity. Its volume increases by a factor of a billion. In the process the star changes colour, because the outer layers are so far from the hot central regions that they cool down. The star becomes a *red giant*.

Despite their low surface temperature, red giants are extremely luminous because they are so gigantic in size. The pantheon of the brightest stars visible to the naked eye is full of red giants: Betelgeuse, Aldebaran, Arcturus and Antares to name a few. The Sun itself will become a red 'monster' in 5 or 6 billion years. When core hydrogen has been burned, the Sun will start to expand; the small planet Mercury, 60 million kilometres away will be vaporised, the atmosphere of Venus will be blown away and the Earth's oceans will boil. Then the Sun will expand even further and consume the Earth. The maximum radius of the future Sun in its red giant phase will exceed the 150 million kilometres of the Earth's orbit. The charred remains of the Earth will continue to circulate in the torrid but extraordinarily thin atmosphere of the giant Sun: the density of the outer layers of a red giant is much less than the best vacuum available in a terrestrial laboratory.

[7] The heavier the atom the more protons it contains in its nucleus and the higher its charge. Nuclei also contain non-charged particles called neutrons, see Chapter 6.

[8] This is called the 'helium flash'.

5

Ashes and diamonds

The life of a star is far from over once it becomes a red giant; gravity becomes more important than ever. A star's destiny is completely controlled by its mass.[1] More massive stars evolve more rapidly, using up their nuclear reserves more quickly. The nuclear lifetime of the Sun is about 12 billion years, but stars which are 10 times more massive will have a nuclear life which is 1000 times shorter. In addition, what they produce is not the same. The most massive stars produce the heaviest elements, a point I will return to in the next chapter. For the moment let us consider the destiny of more modest stars, like the Sun.

The core of carbon and oxygen created in the red giant phase remains inert for thermonuclear reactions. It is not sufficiently compressed by the weight of the envelope. But activity continues all around it. The layers of hydrogen and helium burn in turn, eating away at the star's reserves bit by bit, reaching out into the envelope in search of fuel. The energy produced by the star during this stage of frugal 'nibbling' is only able to support the weight of the layers intermittently. In its death agony the destabilised star begins to pulsate, this stage lasting several thousand years. What was a model of stability begins to vary recklessly, inflating and deflating like a child's balloon, ejecting a puff of gas at each pulsation. Finally it sheds its envelope, leaving its naked carbon–oxygen core.

[1] At least the destiny of a single star, binary star systems are subjected to other factors, which I will discuss later.

The abandoned gas – the cinders – forms a planetary nebula. The shrivelled stellar corpse has the destiny of a diamond, as a *white dwarf*.

Planetary nebulae

The spectacular gas outflow of a planetary nebula awaits not only the Sun but all *average* sized stars with masses of between 1 and 8 solar masses ($M_\odot$).[2] Smaller stars are so thrifty that they have effectively not evolved since their birth, whilst the larger stars burn at high speed and finish their existence in a titanic explosion.

The first planetary nebula was discovered in 1779 by Antoine Darguier, in the constellation Lyra. He found a body 'as big as Jupiter' and resembling a planet. Other similar stars soon joined the list. William Herschel, the musician and discoverer of Uranus, named this new category of celestial bodies 'planetary nebulae', partly because they were nebulous and partly because he thought they could explain the formation of planets. He was wrong about this, but the word 'planetary' has remained as one of the anomalies of astronomical nomenclature. Even the name 'nebula', although less inaccurate, is only a reflection of the modest performance of the instruments of the time. This was still the age when astronomy was a kind of celestial botany, a spelling out of the contents of a world where man could scarcely read. One of these great botanists was Charles Messier, who was primarily interested in comets (Louis XV nicknamed him the 'Comet Ferret'). In 1781 he produced a catalogue of 103 nebulous stars which resembled comets, but unlike them did not move across the sky. With its help comet hunters no longer confused their quarry with these mysterious blurred and immobile patches.

We know today that Messier's catalogue – still useful to amateur astronomers – is a list of many different objects including planetary nebulae (the one in Lyra was number 57 in Messier's list), distant

[2] From now on, the symbol $M_\odot$ will be used to designate the Sun's mass of 2×10^{33} grams, regarded as the astronomical mass unit.

galaxies composed of billions of stars, interstellar clouds, and star clusters belonging to our own Galaxy.

The artist's pallet

Why are planetary nebulae, the remains of gases ejected from a small dying star, amongst the most spectacular stars? Because their gas intercepts the radiation from the incandescent surface of the central star. A body heated to 20 000 K, does not radiate mostly in the visible, but in the *ultraviolet.* This type of radiation carries more energy than visible light[3] and is capable of exciting the atoms in the nebula. Under the continual bombardment of these photons, the electrons jump to higher orbits, and then fall back to their original orbits, releasing radiation of a characteristic colour: the gas becomes *fluorescent.* Each atom in the gas (hydrogen, carbon, oxygen) absorbs ultraviolet radiation and re-emits the energy in other wavelengths whose colour is a kind of signature of the element.

In the internal regions of the nebula, closest to the central star and therefore most exposed to its ultraviolet radiation, oxygen and nitrogen are excited and radiate their characteristic colour, which is green. In the outer regions the ultraviolet radiation is weakened by absorption and can only excite hydrogen, which re-emits it as red light.

A planetary nebula is a star which evolves quickly. Its maximum diameter never exceeds one light year. Its gas, which expands at a rate of 10 to 30 km/s, ends as totally diluted in interstellar space in less than 100 000 years. This is such a brief time on the astronomical scale that one can estimate that a total of between only 20 000 and 50 000 planetary nebulae exist in the Galaxy, formed at a rate of 1 or 2 per year. Of these, only about 1000 are visible, the rest being hidden by the dust of the galactic disc.

[3] See Table 1, page 13.

The garden of white dwarfs

'Strange objects, which persist in showing a type of spectrum out of keeping with their luminosity, may ultimately teach us more than a host which radiate according to the rule.'
Arthur Eddington (1922)

The ashes of planetary nebulae are interesting to astronomers because they sow the interstellar medium with carbon, nitrogen and oxygen. The fate of the surviving star is even more fascinating, as much for observational reasons as for theoretical ones.

After the enormous expansion of the red giant phase and the irreversible decline in the rate of thermonuclear reactions, the star blows off gas and contracts down to the size of the Earth: several thousand kilometres in diameter. Concentrated into this considerably reduced area the temperature rises so much that the star becomes literally *white hot.* These two characteristics of tiny size and high surface temperature give the star the name *white dwarf.*

White dwarfs first appeared in astronomy in 1834, when Friedrich Bessel made a detailed study of the proper motion of Sirius, the brightest star in the sky. Superimposed on its slow movement around the centre of the Galaxy were slight periodic perturbations showing that Sirius belonged to a *binary* system, the mass of its companion being similar to that of the Sun. At this distance, a partner similar to the Sun should have been visible, but was not. The mysterious star, named Sirius B, was only discovered 30 years later by Alvan Clarke. Its luminosity, 10 000 times less than that of its partner, is like a candle flame hidden by a dazzling light.

With such a modest luminosity, it was assumed that the surface temperature of Sirius B would be quite low. In 1917 Walter Adams took its spectrum and found it white (indicating a temperature of about 8000 K), instead of the red which was expected (about 1300 K). How are we to reconcile low luminosity and high temperature? By remembering that a star's luminosity depends not only on its temperature, but also on its size. The most plausible explanation for the dimness of Sirius B is that it has an extremely small stellar radius, of the same order as that of the Earth.

Here is a situation typical of scientific research (which makes it all the more exciting): as soon as one problem is solved, others previously unknown appear. For the companion of Sirius, the luminosity problem was solved by postulating a star the size of a planet. However, a star as small as a planet but as massive as the Sun must have an average density of 800 kg/cm^3, or 40 000 times that of the densest metals known on Earth, such as gold or platinum. To achieve the same concentration of matter in the laboratory would require us to compress the Eiffel tower into a 30 centimetre cube.

These numbers were so surprising for physicists in the 1920s that Arthur Eddington himself called them 'absurd'. However, they were fact, and theory must conform to observational evidence, especially since Sirius B was not the first star to be discovered which differed from the norm: the companion star of 40 Eridan had already been shown to have a surface temperature which bore no relation to its luminosity. During the years that followed the list of white dwarfs grew rapidly, and it became urgently necessary to answer the question: what are white dwarfs made of?

Degenerate matter

Until the beginning of the twentieth century, physicists had never imagined that states of matter could exist which were more dense than those which could be observed on Earth. Water, rock, wood, the human body all have densities within the same order of magnitude: a few grams per cubic centimetre. It was only with the development of the theory of quantum mechanics that scientists were able to understand why ordinary matter has this property.

In an atom, the negative electrons are bound to the positive nucleus by forces of electrical attraction and constantly orbit around it. Just as the repeated shocks of gas molecules hitting the walls of a container generate a pressure, so the electrons bound to a nucleus are responsible for a pressure which prevents matter from contracting beyond a certain limit. This limit is determined by the *Exclusion Principle* discovered by Wolfgang Pauli in 1925.

In pictorial terms this fundamental principle of physics establishes the existence of elementary cells which can contain a maximum of two inhabitants. In 'ordinary' matter (whose density is similar to that of water), most of these cells are unoccupied. It is for this reason that we can say that there is a lot of vacuum in matter: each atom consists of a core which contains most of the mass, surrounded by electrons moving on such distant orbits that if the nucleus was the size of a marble, the atom would measure 2 kilometres across.

However, at the same time as explaining a property of matter which has long been observed, quantum mechanics predicts the possible existence of so-called *degenerate* states of matter, characterised by the fact that all the elementary cells are occupied by particles.

Not all types of matter can become degenerate. Elementary particles are divided into two categories with different collective behaviour at high density or very low temperature: *fermions* (named after the Italian physicist, Enrico Fermi) and *bosons* (named after the Indian physicist, Satyendra Bose, who collaborated with Einstein on the subject). The important characteristic which differentiates between these two large classes of elementary particles is their *spin*. Spin is an intrinsic property of an elementary particle associated with its angular momentum.[4] One of the important things that quantum mechanics revealed is that spin is *quantised*, that is, it can take only certain discrete values, integer or half-integer multiples of a fundamental constant called the 'normalised Planck's constant', $\hbar$ (read h bar). In our daily life, the discrete values of spin pass completely unnoticed because $\hbar$ is so tiny that macroscopic objects have a gigantic spin. The spin of a simple child's top is as great as $10^{30}\hbar$. Thus it is only on the atomic scale that the discontinuity of spin becomes noticeable, along with that of the other quantised physical variables such as energy.

The difference between fermions and bosons is that fermions have half-integer spins ($\frac{1}{2}\hbar$, $\frac{3}{2}\hbar$ and so on), while bosons have integer spins ($0\hbar$, $1\hbar$, $2\hbar$ and so on). The fundamental components of atoms, protons, neutrons and electrons, are fermions with $\frac{1}{2}\hbar$ spin.

[4] Roughly, the product of its radius and its rotational velocity.

The photon (a light particle) is a boson with $1\hbar$ spin. Pauli demonstrated a fundamental principle: *two identical fermions cannot be found in the same quantum state* (this rule does not apply to bosons). This very important law rules out very tightly packed groups of fermions. Let us see in more detail how it works.

In an atom, the quantum state of an electron is defined by its energy (a function of the orbit in which the electron is found) and by the orientation of its spin. This can have one of two directions, either 'up' or 'down', depending on whether it spins in the same or in the opposite sense as its orbit. From Pauli's Exclusion Principle one can deduce that an orbit of given energy can be occupied by two electrons which have two opposite spin orientations. The presence of any further electron in the same orbit is forbidden by nature.

Let us now consider an electron gas in a box. An electron's quantum state is defined by its energy, its linear momentum[5] and its spin. According to quantum mechanics, energy and momentum are also 'quantised' parameters and can take only discrete values. Therefore if electrons are placed in a smaller and smaller volume there will come a point when all the energy and momentum levels are occupied by electrons having all the possible spin orientations. The Exclusion Principle then comes into play and prevents the volume from being populated any more. Consequently the electrons resist any further attempts at decreasing the volume by exerting a colossal internal 'quantum' pressure, called *degeneracy pressure*. The characteristic property of this pressure is its independence of temperature, unlike ordinary gas pressure which increases in proportion to the gas temperature.

White dwarfs unveiled

'I was under pressure, but I could take it!'

World tennis star

The English scientist Ralph Fowler was the first to apply quantum mechanics to astrophysics. In 1925, he suggested that the

[5] The product of its mass and its velocity.

gravitational compression of a star not subject to the pressure of internal radiation was capable of forcing all the electrons into occupying all the possible states, and that the collapse of white dwarfs could be stopped by the degeneracy pressure of the electrons.

Shortly after this, William Anderson showed that for densities exceeding 1 tonne per cubic centimetre the velocity of the electrons approaches the velocity of light; in this case, the electrons are *relativistic*, indicating that their motion no longer obeys Galilean mechanics but Special Relativity. From quantum mechanics, we know that for a given density, relativistic particles exert less pressure than slow particles. It is for this fundamental reason that white dwarfs cannot be arbitrarily massive.

The Indian astrophysicist Subrahmanyan Chandrasekhar was responsible for this important discovery which was to revolutionise theoretical astrophysics. In a famous article in 1931, he proved that white dwarfs had a maximum allowed mass, and calculated this to be 1.4 $M_\odot$. This solution caused a lively controversy. Eddington called it absurd; Chandrasekhar's result[6] meant that the fate of a star much more massive than the Sun became mysterious. But Chandrasekhar was right. One estimates today that stars of up to 8 $M_\odot$ at birth nevertheless form white dwarfs of 1.4 $M_\odot$, because during their lives they lose so much gas in the form of a stellar wind, that their mass is reduced to a value below Chandrasekhar's maximum value. The fate of more massive stars will be clarified after the discussion of white dwarfs when we discuss the theory predicting *neutron stars* and *black holes*.

Hot and cold

White dwarfs, the endpoint of the evolution of moderate-mass stars, are found throughout our Galaxy. One estimates that they currently represent 10% of all stars (about 10 billion), and this proportion can only increase with time.

[6] Chandrasekhar was also the author of a number of important papers on the internal structure of these extraordinary stars. He later tackled a number of other problems in astrophysics with the same success and was awarded the Nobel prize in 1983.

Of these billions of stars, only a few thousand have been recorded. Their luminosity is so low that only the closest white dwarfs have been detected. One of the methods used to discover isolated white dwarfs in fact consists of studying stars of large proper motion – hence nearby – and then taking their spectra to determine their colours. Their positions on the luminosity–colour diagram (see Appendix 1) indicate unambiguously whether they are white dwarfs or low-mass stars.

Let us now take a closer look at a white dwarf. The more massive it is (up to the limit of 1.4 $M_{\odot}$), the smaller its radius; because gravity favours the contraction and compression of degenerate matter. Inside a white dwarf atomic structures are destroyed, and the electrons, freed from the nuclei, can move freely within the 'degenerate sea'. Despite the extremely close packing of the electrons, there is still plenty of space; the nuclei remain so far apart, by comparison with their size, that they behave like molecules of air.

The physical structure of a white dwarf depends essentially on the behaviour of the sea of electrons, whereas the thermal structure depends on the motion of the nuclei. Since degenerate electrons are excellent conductors of heat, the interior of a white dwarf resembles a piece of incandescent metal. The internal temperature reaches 100 million K for newly formed white dwarfs and falls to several million K for old ones. The thermal energy, although associated with a very high temperature, remains much smaller than the electron rest mass energy. This shows that the temperature plays a negligible role in maintaining a white dwarf's equilibrium. In fact, although hotter than the Sun, a white dwarf can be correctly modelled as if its temperature were exactly zero (see Appendix 2).

The interior of a white dwarf is protected from the cold of interstellar space by a thin layer, several kilometres deep, which is very opaque and strongly insulating, consisting of non-degenerate matter whose temperature is less than 100 000 K. It is this surface temperature, 10 times the Sun's, which is responsible for the luminosity; but as the emitting area is so small, the total luminosity is low and white dwarfs are pale phantoms very difficult to detect at great distances.

The age of crystal

Since there are no thermonuclear reactions to provide it with new energy, the white dwarf cools down at the rate at which it emits radiation. However, the white dwarf is by nature thrifty. Once formed, a white dwarf takes billions of years to cool down. Initially, the non-degenerate nuclei move freely, like the molecules of an ordinary gas, and their kinetic energy is responsible for the temperature. Eventually, there is a critical moment when so much kinetic energy has been gradually lost through radiation that the remainder is less than the electrostatic energy of the nuclei, tending to imprison all the nuclei in a rigid framework. The motion begins to slow down and the nuclei arrange themselves into a *crystalline lattice*, while the degenerate electrons continue to move freely within it. The aged white dwarf practically stops emitting radiation and transforms itself into a giant crystal harder than a diamond. It has become a *black dwarf*.

The darkening of a dwarf is so slow that it is possible that no black dwarf has been formed since the beginning of the Universe and the appearance of the first stars, 15 billion years ago. A lot of patience is required. The Sun, at present halfway through its main sequence will take another 5 billion years to reach its 'old age' as a planetary nebula. It will be active briefly for another 100 000 years, will slowly die away as a white dwarf for another 10 billion years, and will finally pass gently into an eternal age of crystal.

Future glitter

Single stars like the Sun are in a minority. More than half the stars in the Galaxy are in binaries. Some of them even maintain very close gravitational relationships with two, three or even four partners. The white dwarf Sirius B has a companion, which, however, is too distant to influence its destiny. As an isolated white dwarf, Sirius B is probably destined for inexorable cooling. However, if the two stars are closer together, the long-term evolution of a white dwarf can be transformed.

The main reason for this difference is *mass transfer* between the two partners. The companion of a white dwarf, if it is either very close or in a phase of large spatial expansion (red giant) can have its envelope progressively removed by the white dwarf. Generally speaking, the pulled-off gas does not fall directly onto the surface of the white dwarf because of the centrifugal forces caused by its orbital motion. It accumulates instead as a more or less flat structure in orbit around the white dwarf, called an *accretion disc* (Figure 19). The impact on the disc of the flow of gas from the companion star produces very strong local heating, a *hot spot*, which can shine like a star and indirectly reveal the presence of the white dwarf. In other cases, particularly if the white dwarf is highly magnetised, the disc is unable to form and the gas is channelled along the magnetic field lines in the direction of the white dwarf's polar caps. On arrival its impact produces highly variable optical, ultraviolet or even X-ray radiation which erratically lights up the white dwarf. The white dwarf then becomes visible as a *cataclysmic variable*.

This relatively stable arrangement is interrupted by periods of intense and sudden activity, and the result is a *nova*, or 'new star'. Originally the name designated the class of stars whose luminosity increased suddenly and died away slowly. In fact, novae include various categories involving a wide range of phenomena, but they all involve a dense star in a binary system.

The mechanism of a nova is in all likelihood a *surface thermonuclear explosion*. The gas settles continuously on a white dwarf, where it is compressed and heated by the gravitational field. At a certain critical stage, the hydrogen – its main constituent – suddenly fuses and the outer layer of the white dwarf explodes. For a period of several weeks, the white dwarf shines brightly, revealing its presence as far away as the edge of the Galaxy.

Some novae are recurrent, that is, the explosions repeat at intervals of a few months. Other novae explode just once and release a lot more energy. One of the most brilliant ever observed was Nova Cygni 1975, which shone for three days with the brightness of a million suns. This correlation between the intensity of the explosion and the period of recurrence confirms the model of mass transfer between partners; the energy liberated is a measure of the amount of gas that has accumulated on the surface of the white dwarf.

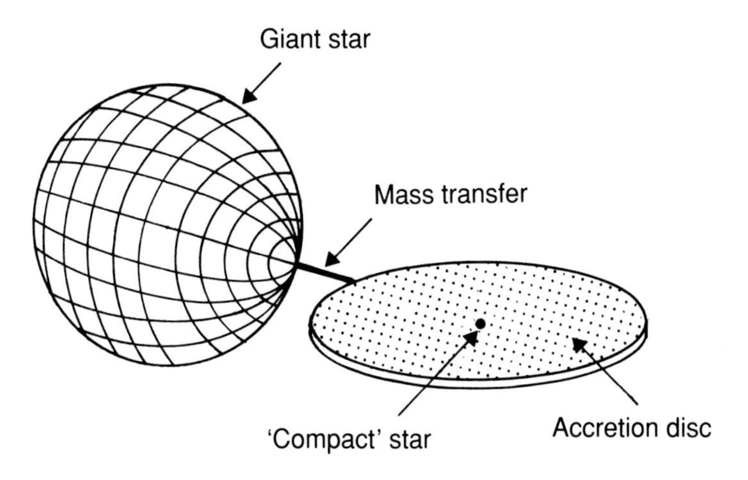

Figure 19. An accretion disc in a binary system consisting of a giant star and a compact star.

The mechanism of mass transfer between a 'normal' star and a 'compact' companion plays a crucial role in most very energetic astronomical phenomena. I will return to this point in more detail in Part 4, because close binary star systems can sometimes provide spectacular evidence of the presence of a black hole which would otherwise remain perfectly invisible.

6

Supernovae

The nuclear ladder

Nature's list of elements does not consist solely of hydrogen, helium, carbon and oxygen. Living matter, wood, earth or rock also need some silicon, magnesium, phosphorus, sulphur, iron and other 'heavy' atoms, with nuclei consisting of more than twenty protons and neutrons. Where are such elements forged, if the Sun and most stars fail?

It is in stars again, but only a tiny fraction of the total number: the most massive. Only stars over 8 $M_{\odot}$ when they leave the Main Sequence are able to make heavy nuclei. The crucible is the core of the star, crushed by the weight of the envelope. The raw materials are the 'ashes' of hydrogen and helium, that is, carbon and oxygen. The process begins when the core temperature rises above 600 million K.

At this temperature carbon cannot hold out; the nuclei crash into each other, fusing to form neon and magnesium. A production line is set up as each new thermonuclear reaction releases more energy, raising the temperature and permitting more transmutations. At a billion K, neon can carry off a helium nucleus to form magnesium. At 1.5 billion K oxygen also starts to 'burn', producing a range of heavier nuclei: sulphur, silicon and phosphorus. At 3 billion K silicon burns, setting off several hundred nuclear reactions which heat the furnace still more, and so on. In the welter of several thousand reactions, still heavier and richer nuclei are forged. The last stages of the life of a star intensify to a paroxysm: the heavier a nucleus, the shorter its combustion time. For a 'model' star of 25 $M_{\odot}$, carbon burning lasts 600 years, that of neon a year, oxygen 6 months and silicon a day.

The giant onion

Nuclear transmutation cannot continue indefinitely at such a rate. The 'flood' of transmuted elements converges towards a single nuclear species: *iron*. This is a very special nucleus, because the 56 protons and neutrons in the nucleus are so tightly bound together that no fusion energy can tear them apart. Iron constitutes the ultimate ashes of the core of a massive star.

The star is now composed of a core inert to thermonuclear reactions surrounded by layers which burn in succession. The star has constantly to readjust its equilibrium by expanding its envelope. It dilates to an enormous size and becomes a *red supergiant*.

The red supergiants are the largest stars in the Universe. If one was placed at the centre of the Solar System, it would engulf all the orbiting planets out to Pluto, 5 billion kilometres away. The internal structure of a red supergiant is sometimes described as being like an *onion*, because it consists of concentric layers burning different chemical elements (Figure 20). The lightest elements burn in the outer layers, where the temperature is lowest, and the heavier elements burn in the inner layers around the inert iron core.

Neutronisation

Although its temperature is greater than a billion K, no more energy flows from the iron core. It is 'cold' and no longer able to maintain the gravitational equilibrium of the supergiant by itself. The core becomes more dense and the electrons become degenerate. There is a lull in the star's activity while the enormous pressure of the degenerate electrons is momentarily able to support the weight of the envelope.

But we recall that a cold mass of degenerate electrons is incapable of supporting more than 1.4 $M_{\odot}$. This is the Chandrasekhar limit above which no balance is possible between gravitation and the quantum pressure of the electrons. Fresh iron is constantly being made in the layers surrounding the supergiant's core. Being heavy, it sinks and joins the core. Eventually the fatal moment arrives when the central mass of iron nuclei and degenerate electrons exceeds the Chandrasekhar limit.

One can estimate that all stars greater than 10 $M_\odot$ (core and layers included) are able to develop a core more massive than 1.4 $M_\odot$. The density at this stage reaches a billion g/cm^3. The core of degenerate matter suddenly gives way and collapses. In a tenth of a second, the temperature soars to 5 billion K. The photons which flood out transport so much energy that they explode the iron nuclei, reducing them to a dust of helium nuclei. This process is called *photodisintegration*.

Unlike nuclear fusion reactions, which increase the size of the nuclei and release energy, photodisintegration *breaks up* nuclei and *absorbs* energy. Nothing more catastrophic could happen for the equilibrium of the stellar core; less and less able to withstand the relentless crushing force, its temperature continues to increase until the helium nuclei themselves disintegrate into their basic constituents: protons, neutrons and electrons. But at these temperatures, the velocities of the electrons approach that of light. Thus, although degenerate they become even less capable of resisting the compression forces. In a tenth of a second they are pushed into the interior of the protons themselves. Their electric charges are neutralised. *Neutrons* are created, amid a huge burst of *neutrinos*.

The neutrino (which means 'little neutral one') is an elementary particle whose existence was predicted in 1931 by Pauli before being detected experimentally in 1956. Normally, the neutrino has so few interactions with matter that it can travel great distances without being stopped or deviated from its course. But in the imploding core of a massive star the deluge of neutrinos liberated by neutronisation produces so much energy that the stellar envelope feels the shock and absorbs a substantial part of the neutrinos. The rest escape from the star at the speed of light and cross interstellar space without the slightest resistance.

The neutron is a constituent of atomic nuclei (a *nucleon*), along with the proton. It was discovered only in 1932, because it cannot

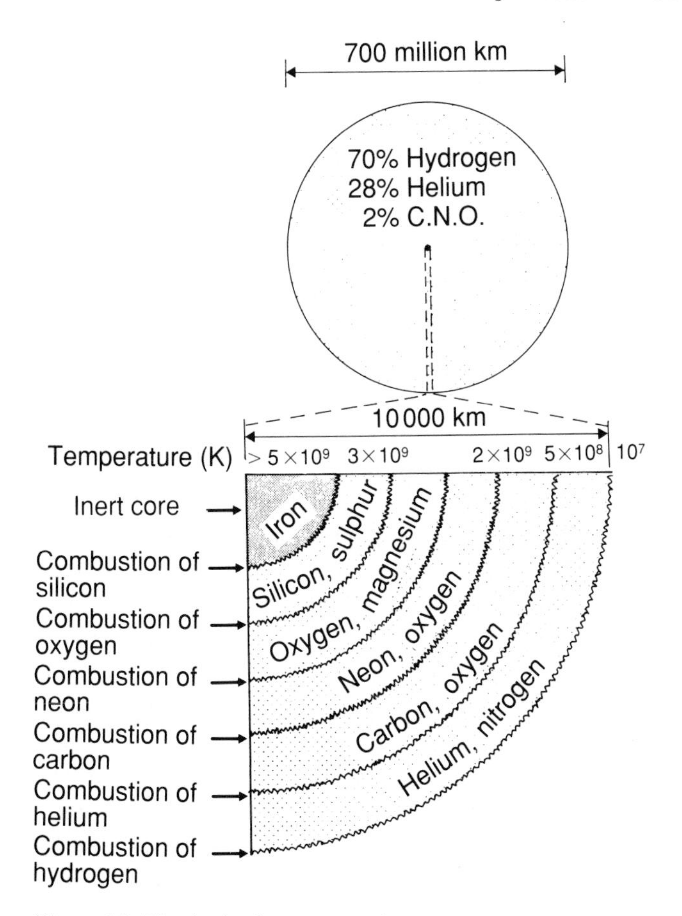

Figure 20. The 'onion' structure of a massive supergiant star just before it explodes in a supernova. The chemical composition of each layer is the product of thermonuclear reactions occurring in various regions of temperature and density as these increase towards the centre of the star.

survive alone. Once it is separated from a nucleus, it has a very short lifetime; after about 10 minutes it disintegrates spontaneously, losing its electrical neutrality and generating a proton, an electron and an antineutrino.[1]

[1] The antineutrino is the neutrino's antiparticle. The disintegration of a free neutron is the inverse reaction of electron capture by a proton, which occurs in the cores of collapsing stars.

Now we reach the most important point: since its spin is a half integer, it is a *fermion*, obeying the Pauli Exclusion Principle like the electron. Its 'occupation volume', however, is much smaller: the separation between two neutrons can be as little as 10^{-13} centimetre, i.e. the neutrons can touch each other. Thus neutronisation is accompanied by an *implosion* of matter and an enormous increase in density towards a degenerate state. A quarter of a second after the star begins to collapse a density of 10^{14} g/cm^3 is reached (equivalent to 100 million tonnes in one thimble). This is the density of atomic nuclei, as if all the electrons had been removed from ordinary matter and the atomic nuclei allowed to touch. There is no more 'empty space' left in the stellar core, so it has become a sort of giant atomic nucleus consisting mainly of neutrons. This new degenerate state of matter, which is much more dense than a white dwarf, is called a *neutron star.*

Blow out

Once matter reaches nuclear density, it becomes virtually impossible to compress it any further. The non-neutronised outer layers of the star which fall onto the core's surface at speeds of about 40 000 km/s encounter an incredibly hard wall. They are stopped dead and rebound as a *shock wave.*[2]

In the gravitational collapse following the rebound of the envelope, the shock wave propagates from the centre outwards and reaches the star's surface after several days. It carries with it an enormous amount of energy, and literally blows the envelope radially outward. Our 'model star' of 25 $M_\odot$ would eject a mass of 24 $M_\odot$ leaving a residue of 1 $M_\odot$ as a neutron star. This phenomenon is called a *supernova.*

It is difficult for us to appreciate the cataclysmic nature of a

[2] A shock wave is the propagation through a material medium of a discontinuity front which causes an abrupt modification of certain physical properties such as pressure, temperature and density. In nature, all explosive situations are accompanied by shock waves, which are always produced when the velocity of the expansion of matter becomes greater than the local velocity of sound. Breaking the 'sound barrier' occurs when an aeroplane's velocity exceeds 330 m/s, and is accompanied by a shock wave which propagates through the atmospheric layers and produces an acoustic 'bang'.

supernova explosion. In a few days it spews out as much energy as a star does over hundreds of millions of years during its period in the Main Sequence. Its luminosity increases several billion times, so that it seems that for a few days the 'new' star is as bright as a whole galaxy.

In comparison, the phenomenon of a planetary nebula accompanying the condensation of a star into a white dwarf seems a very peaceful death, a sort of second class burial; the explosion of a supernova is a violent death, ejecting more ashes and leaving a denser stellar corpse.

The interstellar medium is enriched by the heavy elements formed in the 'onion': the gases blown out by supernovae play an even more important role in the evolution of the galaxies than those of the planetary nebulae. The vast molecular clouds in which entire generations of stars are born are sown by the explosions of nearby supernovae. Five billion years ago, when the Sun and its associated asteroids, meteors, comets and planets were released from the primordial cloud, the Galaxy was already 10 billion years old and a number of massive stars had already burnt away, distributing their ashes throughout the Galaxy. The Earth received its complement of heavy elements from the nuclei of stars which had long since disappeared.

Observing supernovae

The phenomenon of a supernova is of course not limited to massive stars in our Galaxy. But since apparent luminosity diminishes rapidly with distance, it took the advent of large telescopes in the twentieth century before we were able to observe supernovae exploding in other galaxies. Today, several hundred supernovae have been catalogued, appearing at an average rate of two per month, scattered throughout the several hundred thousand neighbouring galaxies. One can deduce that the frequency of explosions in a given galaxy is about four per century.

With the naked eye we see only the stars in our Galaxy. Astronomical observations have been recorded in writing for the past 2000 years. During this period, about 100 supernovae have exploded, but only a few of them have been recorded.

The main reason for this rarity is that the Sun lies in the galactic disc, actually where the majority of the massive stars producing supernovae are found. But the optical penetration of the galactic disc (visible in the night sky as the Milky Way) is greatly reduced by vast amounts of dust which absorb the visible light. The galactic disc can be probed only up to a distance of several hundred light years, so that we have access to only a tiny fraction of the interesting volume of our Galaxy.[3]

Recent and future developments in observational astronomy should be able to overcome this problem. During the explosion of a supernova, not only are light photons emitted, but other radiation too, which is able to move without being obstructed by cosmic dust. In particular, *neutrinos* are emitted in profusion, and are capable of travelling many light years without interacting with matter. If they could be detected on Earth we would have a mine of entirely new information about the sources which emitted them. The problem is indeed how to detect them; since they barely interact with matter, they do not interact readily with normal measuring instruments.

The thermonuclear reactions occurring in the core of the Sun produce a constant stream of neutrinos, of which an infinitesimal fraction is detected at Earth by a vast reservoir containing 600 tonnes of carbon tetrachloride (CCl_4) buried in a gold mine in South Dakota. A neutrino hitting a chlorine atom in this weird swimming pool transforms it into argon, which can be extracted from the mixture. (A more recent European version uses gallium as the target.) Supernovae neutrinos are more energetic than solar neutrinos. Detectors for these were actually designed for other purposes. Particle physicists have constructed giant water tanks underground (thus sheltered from cosmic rays) so as to detect proton decay and the resulting flashes of light. The question of a finite proton lifetime, raised by recent unified theories of particle interactions, is important since the proton is the fundamental constituent of atomic nuclei. So far there has been no detection of a decaying proton. On the other hand these watery detectors are

[3] We will see later that radio, infrared and X-ray radiation is less absorbed and can reach the Earth.

sensitive to energetic antineutrinos, such as are emitted by a nearby supernova; an antineutrino interacting with a proton in the water tank produces a neutron and a positron (antielectron). This emits a flash of Cerenkov radiation which is registered by some of the thousands of photocathodes immersed in the tank. This technique achieved a triumphant success in February 1987 during the appearance of supernova SN 1987A, which we shall return to later.

Another type of radiation emitted by supernovae is perhaps even more promising: it is not electromagnetic or neutrino radiation, but gravitational.[4] Since Einstein's theory of General Relativity predicts the propagation of waves of curvature when the gravitational field varies rapidly, such waves should be produced when a star collapses. Towards the year 2000, gravitational telescopes should be able to detect the signals emitted by supernovae exploding up to 100 million light years away. At these distances several thousand galaxies are accessible and the telescopes should detect one burst of gravitational radiation per month.

Historical supernovae

Although we dream of tomorrow's astronomy, we are not condemned to wait patiently for a star in its death throes. The astronomy of the past is a mine of priceless information: humanity's written records contain astronomical treasures only waiting to be used.

The cataclysmic deaths of massive stars left traces in the annals of observational astronomy long before the invention of telescopes. In the Far East, the professional astronomers – usually astrologers – were employed by rulers to watch the skies and to report and interpret unexpected events. Many of these events were recorded in remarkably detailed works throughout several Chinese dynasties; some of these works have survived from as long ago as 200 BC. Older works have unfortunately perished. This irreparable loss resulted from the overweening pride of a single man, Ch'in Shih-Huang, who declared himself to be the first 'true' emperor of

[4] The question of gravitational waves will be discussed in more detail in Chapter 18.

China. Deciding that the history of the world began with his reign, he decreed in 213 BC an orgy of destruction in which most of the older documents were lost.

Fortunately, China was not the only country with a keen interest in astronomy. Both Japan and Korea had been making regular astronomical observations since 1000 BC. From this date, it is possible to find simultaneous records of the same event, enabling scientists to authenticate phenomena often cryptically described.

The exact number of historically recorded supernovae remains uncertain, but it is no more than about 10. However, not all the records were compiled by historians interested in astronomy, even fewer by astronomer historians knowing oriental languages.

The first three new stars observed by the Chinese get a very brief mention. One was visible for 20 months in Centaurus in AD 185; another, in AD 396, shone for 8 months in Scorpio; the third, in AD 827, also appeared in Scorpio.

In AD 1006, a supernova in Lupus was recorded by enough different sources for its authenticity to be verifiable. The new star was observed by the Europeans (and recorded in the monasteries of medieval Europe), by the Arabs, the Chinese and the Japanese. It remained visible to the naked eye for 25 months, and, according to a description from Iraq, its brilliance exceeded that of a quarter of the Moon.

Identification of a star

'I bow low. I have observed the apparition of a guest star. Its colour was an iridescent yellow [. . .]. The land will know great prosperity.'

Yang Wei-T'e, Imperial astronomer (1054)

The most famous supernova in history (at least to us) was observed in 1054 by the Japanese and Chinese. The most meticulous description of the event was recorded in China by Yang Wei-T'e, the Imperial Astronomer to the Chinese court of the Sung dynasty, who knew the constellations well. On the day of Ch'ih Ch'iu of the fifth Moon of the first year of the period of Shih-Huo

– or 4 July 1054! – Yang Wei-T'e noticed the apparition of a strange star in the sky. A few minutes before sunrise, an unknown star rose above the horizon, much brighter than Venus or any star ever seen in the sky. The Imperial Astronomer called it a 'guest star' and noted its appearance in the records. He reported it to his master and had the good sense to interpret its arrival as a favourable omen, and then maintained a careful watch over it. The guest star remained visible *in broad daylight* for 23 days, and could be seen in the night sky for two years. It finally disappeared and the spectacle was over. Yang Wei-T'e had witnessed the explosion of a supernova which was as brilliant as 250 million Suns.[5]

This was all forgotten until John Bevis, an English amateur astronomer, discovered a nebula in Taurus in 1731. As a diffuse object, it appeared as number 1 in Messier's celebrated catalogue. Lord Ross, who studied its shape, christened it the 'Crab Nebula' in 1844. In 1919, thanks to a translation of the Chinese annals, the Swedish astronomer Lundmark realised for the first time the connection between the Crab Nebula and the supernova of 1054. Finally in 1928 Edwin Hubble, the father of modern cosmology, measured the expansion velocity of the Crab Nebula, and by extrapolating backwards was able to estimate its age as about 900 years, which agreed with the date of the explosion in 1054. The connection between the exploding star and its gaseous residue was no longer in any doubt.

The Renaissance supernovae

In 1572, a supernova was observed in the West by the Danish astronomer Tycho Brahe, in the constellation Cassiopeia. For several days it shone as brightly as Venus. As the first supernova to be scientifically examined it played a very important historical role. At the time, the Greek and Arab view that the Earth was the centre of the Universe and the stars were all fixed to a distant sphere was still generally held. Tycho Brahe showed that the new star of 1572 must be further away than the Moon and

[5] Because of the distance of the star, the explosion actually took place 5000 years earlier.

therefore in the sphere of the fixed stars. He thus shook to its foundations the theory of the immutability of the stars, already cast in doubt by Copernican theory, thus preparing the ground for Johannes Kepler's great astronomical revolution.

The 1572 explosion was also responsible for the subsequent name 'supernova' given to such events in the twentieth century. It was as if an ordinary new star (a nova) was situated only tens of light years away. However, at such a short distance a white dwarf – the residue of a nova – should have been observable by telescope, but was not. So the new star in 1572 had to have been much brighter than a nova and much further away. For this reason Fritz Zwicky and Walter Baade suggested the name supernova in 1937.

The supernova of 1604 was simultaneously observed in Europe, China and Korea. It is often referred to as Kepler's supernova, because it was the famous German astronomer who determined its exact position. In 1943, Walter Baade discovered nebulosity around the position of the explosion.

The list of supernovae recorded in our Galaxy ends here.[6] The last one was four centuries ago. But the unexpected explosion in February 1987 of a supernova not in our Galaxy but just outside, i.e. in the Large Magellanic Cloud, had a considerable effect and monopolised the attention of observational and theoretical astronomers for many months of feverish activity. We will discuss it in the last section of this chapter.

The remains of the feast

'Show me what you leave on your plate and I will tell you who you are.'

French proverb

Whilst the increase in the brightness of a supernova only lasts a few months, the residues ejected by the explosion and blown into interstellar space can be observed for much longer. Thus the gaseous debris of supernovae which exploded in the distant past is still observable today. However, the remains of supernovae are

[6] Except perhaps for Cassiopeia A (see below).

relatively short-lived. Some of them are so dispersed and weak that their visible light no longer reaches us. However, as they expand, they collide with the interstellar medium and produce radio waves and X-rays. About 20 remnants are observable in the optical spectrum and more than 100 in the radio.

The most famous supernova is the Crab Nebula, born in the 1054 explosion. However, the Vela supernova remnant in the Gum nebula must have resulted from an explosion in about 9000 BC, when men surely watched the sky but did not record their observations. At maximum the star must have been as bright as the first quarter of the Moon. The explosion of the magnificent Cygnus Loop must have occurred between 20 000 and 30 000 years ago.

The remains of the supernovae are full of information on the nature of the explosion which caused them. Supernovae can be classified into two types depending on their luminosity evolution. In Type I the maximum luminosity is greater than Type II, and its decrease more irregular, by stages.

Theoretical astrophysicists still debate on the interpretation of these two types of supernovae. Some of them, comparing the spectra of the two types of supernovae, believe there is simply a difference in the chemical composition of the exploding stars. Stars belong to one of two 'Populations' depending on their chemical composition and age. Population II consists of old stars which were present at the formation of the galaxies and for this reason contain few 'metals'.[7] This type of star dominates elliptical galaxies which have been stripped of much of their gas and where no new stars are formed, and the halos of spiral galaxies. Population I consists of young stars, formed in the discs of spiral galaxies and enriched at their birth with 'metals' made by previous generations of stars. Now Type I supernovae are observed in both spiral and elliptical galaxies whereas Type II supernovae are only observed in spiral galaxies. It is thus tempting to suppose that Type II supernovae occur in Population I stars and Type I supernovae in Population II stars. The correspondence is at best poor, and the situation is probably more complicated.

While the theoreticians agree that Type II supernovae are

[7] For the astrophysicist, a metal is any element that is not hydrogen or helium.

explosions of massive stars (greater than 10 $M_\odot$) and are accompanied by the formation of neutron stars, interpretations of the Type I supernovae are numerous. Models show that the gravitational collapse of an isolated star of between 1 and 8 $M_\odot$ does not amount to much: a planetary nebula and a white dwarf or possibly a neutron star with a low energy release. On the other hand, stars between 8 and 10 $M_\odot$ can explode as Type I supernovae, the energy being supplied by the combustion of carbon.

Dangerous liaisons

A different explanation, currently in vogue, invokes a completely different explosion mechanism: Type I supernovae involve white dwarfs composed of carbon and oxygen which are members of a close *binary* system. In this model helium is extracted from the companion and slowly deposited on the surface of the white dwarf; when the layer reaches a critical temperature and density, helium fusion is triggered, causing a flash then a gradual decrease in the luminosity like that observed in the Type I supernovae. In this interpretation 'supernovae' for once truly merit their name, being in effect huge novae.[8]

A variant of this binary model assumes that the white dwarf is near to the 1.4 $M_\odot$ stability limit. Under these conditions, the continuous deposition of gas onto the surface increases its mass until it crosses the fatal threshold. The violation of the Chandrasekhar limit results in a gravitational contraction, slight but sufficient to cause carbon (the main component of a white dwarf) to react and be instantaneously transformed into nickel and iron. The white dwarf is disrupted in the explosion.

Another version of these 'dangerous liaisons' has recently been suggested. In a binary system consisting of two very close white dwarfs, gravitational radiation dissipates the orbital energy of the system, bringing the two white dwarfs closer in a time much shorter than the age of the Universe. It is possible that the collision of the

[8] We recall that novae involve hydrogen burning on the surface of a white dwarf in a binary system.

white dwarfs gives rise to the release of an amount of energy comparable to that of a Type I supernova.

In any case, the proliferation of supernova models gives an idea of the difficulties faced by theoretical astrophysicists trying to explain an extreme state of matter which cannot be reproduced in the laboratory.

Close encounters of the third kind

Studies of the supernova remnant Cassiopeia A have further complicated the way we interpret the different mechanisms of supernovae. This nebula has the advantage of being observable in the optical as well as the X-ray and radio regions. Measurements of the expansion velocity show that the supernova must have exploded in about 1670 at a distance of only 9000 light years. However, no explosion was recorded at this date, although enough astronomers were studying the sky and such a nearby event would not have been missed; for a month, it would have shone more brightly than Sirius. Recently some researchers in the history of science have apparently found a trace of the new star by analysing the celebrated star catalogue (with magnificent engravings of the constellations) of the Astronomer Royal John Flamsteed. This appeared in 1725 but was based on observations carried out in 1680. At the present position of Cassiopeia A it shows a star of 6th magnitude (limit of visibility to the naked eye), called 3 Cassiopeiae by Flamsteed, but missing from earlier catalogues and from the succeeding compilation dating from 1835. No-one at the time, not even Flamsteed, noticed that this feeble star had only recently appeared in the sky.

How do we explain the faintness of the explosion? It is possible that a tremendous amount of dust formed in the expansion envelope could have absorbed all the light from the centre. However, other puzzling facts reduce the credibility of this argument. On the one hand, the absence of iron means that the chemical composition of the nebula is different from the remnants of Type I and Type II supernovae. On the other hand, Cassiopeia A does not appear to have left a remnant neutron star: 300 years after formation the surface temperature of a neutron star would still be about 3 million K and would be a detectable X-ray source. In other words, this may be a third type of supernova, called 'Type III',[9] which is much rarer. It may have been caused by a different mechanism of stellar explosion, triggered not by the collapse of the nucleus but by the instability of an ultra-hot star belonging to the 'Wolf–Rayet' class. A theoretical model, recently developed at the Centre d'Etudes Nucléaires (at Saclay in France) predicts that there is a maximum luminosity which is only 100 million times the Sun's luminosity, that is, 10 times less than a 'normal' supernova. Such an explosion would cause the star to disintegrate completely without leaving a compact residue.

There is yet another hypothesis which is perhaps more seductive: the collapse of a degenerate core occurs, but instead of producing a neutron star, it produces a *black hole*. As will be shown later, this type of star does not have a solid crust, it would therefore not be able to cause the stellar envelope to bounce and so the power of the supernova would be considerably reduced.

The supernova in the Magellanic Cloud

During the night of 23–24 February 1987 the Canadian astronomer Ian Shelton, working at the Las Campanas observatory in Chile, had the extraordinary good fortune to be the first 'professional' to discover a supernova (a night assistant had just noticed it with the naked eye as a 4th magnitude star). The Large Magellanic Cloud, in which the supernova occured, is an irregular galaxy and a satellite of the Milky Way, orbiting at about 170 000 light years. A telegram was sent urgently to the Bureau of the International Astronomical Union, and caused an immediate sensation in the astronomical community.

The supernova, called SN 1987A, was the first visible to the naked eye since that seen by Kepler in 1604, and the closest. But as it was visible only from the southern hemisphere, only observatories in Chile, Australia and South Africa could train their telescopes on it. As night fell in Australia, an astronomer from that country indentified the supernova with a star previously known as

[9] Certain authors prefer to call them Type Ib.

Sanduleak - 69 202, a 12th magnitude blue giant. This posed the first interesting problem for the theoreticians, as they had rather expected it to be the explosion of a red giant. The second puzzle was that the spectrum of the exploding star showed lines of hydrogen, classifying a Type II supernova (explosion of a massive star); but right from the beginning its 'light curve' (variation of luminosity with time) showed anomalies compared with archetypes of this class. In particular, the maximum brightness was almost 100 times less than expected.

Learning of Shelton's discovery, the theoreticians at Princeton immediately set to work and produced in two days an article predicting 'in reverse' that the available neutrino detectors should have caught neutrinos several hours before the luminous apparition of the supernova. They calculated the number and energy of the neutrinos. In a Type II supernova neutrinos are produced by neutronisation, i.e. electron capture by atomic nuclei during the collapse of the core. The neutrinos thus remove most of the energy of the supernova. The neutrino luminosity is equal to the luminous energy liberated in 1 second by 100 million complete galaxies! This fabulous figure corresponds to a flux of 100 billion neutrinos penetrating each square centimetre of the Earth's surface – or of our skins.

Now on 23 February, almost 22 hours before the supernova appeared in visible light, a water detector at the bottom of the Kamioka mine in Japan twinkled 11 times in an interval of 11 seconds under the impact of the burst of antineutrinos from SN 1987A. This result was announced by the Kamioka research team only after 15 days of unremitting labour spent analysing the data. A little later an American team announced a similar result: at the same time as in Japan, 8 flashes had illuminated their detector deep in a mine in Cleveland. If the southern hemisphere had received the light from the supernova, it was the northern hemisphere which detected its neutrinos. Nineteen in all, a tiny harvest, but of major importance: it confirmed not only that SN 1987A was not a Type I supernova (explosion of a white dwarf in a binary system, emitting no neutrinos), but opened a new astronomical era in which not just light but also neutrinos could be detected from stars other than the Sun.

Let us return to the light curve. Its anomaly of the first few days disappeared after a few months: the luminosity followed an exponential decay characteristic of the radioactive disintegration of cobalt 56. This was a new success for the theoretical models: this element is the main product of the explosive nucleosythesis in massive stars. The initial anomaly could be retrospectively explained in terms of the particular nature of the parent star, which had exploded in the blue rather than the red. Sanduleak - 69 202 has probably been a red supergiant, immoderately bloated after helium burning; but had lost its envelope because of a powerful stellar wind, blowing for 10 000 years. This reduced it to a brilliant blue star of smaller size (40 times the Sun's diameter rather than 500 times). Theoreticians were kept busy developing new models and fitting them to the observations, modifying them over the succeeding months as the new data arrived. There remains the most important question of all for us: is the remnant of the explosion a neutron star or a black hole? Both are possible since the parent star had a mass of about 20 times that of the Sun. For four years now various detectors have been trained on the explosion site to discern the traces of a neutron star (a black hole would be 'less interesting' as it would give no detectable signature). Despite a few false alarms these efforts have remained fruitless for the moment. This is not surprising. The remnant is still shrouded in the inner layers of the expanding nebula, but if it is a neutron star it will show itself sooner or later once the last veils have been diluted to transparency. After a few years, or a few decades, the X-rays from the ultra-hot surface of the neutron star will emerge. We might speculate on the birth of a baby radio pulsar if by a lucky chance its beam crosses the line of sight to the Earth (see Chapter 7). We can reasonably hope for an indirect signature, such as heating of the expanding nebula by the central pulsar. Whatever happens, the Magellanic supernova will have been one of the great astronomical events of the century.

7

Pulsars

Science is built from theories and experiments (or in the case of astronomy, observations) which confront each other, victory going to one or the other from time to time. Neutron stars are one of the most beautiful examples of theoretical prediction preceding observational discovery.

Sir James Chadwick discovered the neutron in the laboratory in 1932.[1] It is said that on the same day the eminent Russian physicist Lev Landau and his group speculated on the existence of stars consisting entirely of neutrons. Unfortunately, Landau did not publish his reflections immediately, and it was left to two American astrophysicists who had followed the developments in particle physics to cull the ripe fruit two years later. Inspired by the analogy with white dwarfs, which, as Ralph Fowler had proposed, supported their own weight by electron degeneracy pressure, Fritz Zwicky and Walter Baade suggested that neutrons were able to exert a degeneracy pressure capable of supporting a stellar corpse more massive than the Chandrasekhar limit. The two were very interested in the Crab Nebula, the remains of the 1054 supernova, at the centre of which was a shrivelled up body which was not a white dwarf.

Shortly before the outbreak of the Second World War, Robert Oppenheimer[2] and G. Volkoff developed a proper theory of neutron stars. They showed in particular that hydrostatic

[1] He was awarded the Nobel prize in 1935.
[2] The future 'father' of the atomic bomb.

equilibrium of a degenerate neutron is possible for a star with a mass similar to that of the Sun.

Their work was politely ignored by the astronomical community. The 1955 edition of Camille Flammarion's famous *Astronomie populaire*, through which I first developed a love for astronomy, devoted only a few lines to Zwicky's 'revolutionary' theory: 'It consists of vague ideas which cannot be tested by observation.' The observational test would have to wait 12 years.

Lighthouses in the sky

'Here was I trying to get a PhD out of a new technique, and some silly lot of little green men had to choose my aerial and my frequency to communicate with us!'

Jocelyn Bell

In 1967 Jocelyn Bell, a young graduate student at the University of Cambridge, was given by her supervisor, Anthony Hewish, the task of testing and developing new radio telescopes for measuring emissions from distant radio sources. While analysing by hand hundreds of metres of millimetric chart paper from the recorders she became intrigued by the presence of periodic signals spaced at intervals of precisely 1.337 301 33 seconds. Miss Bell had just accidentally discovered a star emitting radio pulses: a *pulsar*.

Other findings were quickly added to the list. In 1968 pulsars were also discovered in the Crab and Vela supernova remnants. For several months there was great excitement, even spreading beyond the astronomical community. One opinion was that these signals from space arriving at such precise intervals could only be transmitted from an artificial source, deliberately pointed at us by aliens, the 'little green men' so beloved of science fiction. In the absence of an official title, the first pulsars were humorously called LGM1, LGM2 and so on (for 'Little Green Man'). What was merely an astronomical joke caught the imagination of the popular press, excited at the idea of contact with aliens.

At the same time, theoretical astrophysicists considered the question soberly. In 1968 Franco Pacini and Thomas Gold suggested that a pulsar was a rapidly rotating neutron star. The basic idea is as follows. A neutron star is highly magnetised and its magnetic field lines channel electrically charged particles (electrons and protons), enabling it, by 'synchrotron radiation', to emit a beam of radio waves which rotates along with the star. As the star revolves, a pulse is received at the Earth, at the moment when the beam sweeps across the aperture of the radio telescope (Figure 21). This *lighthouse* effect occurs because the rotational axis and the magnetic axis are not aligned, a common phenomenon in astronomy.

This simple and complete explanation was immediately accepted and remains the working model used by specialists. Anthony Hewish received the Nobel prize in 1974 for the design of his radio telescope – and the discovery of pulsars was mentioned in an appendix to Jocelyn Bell's thesis!

A more extreme type of star

Why are the rotation and the magnetic field so important in a pulsar?

A neutron star is formed when the core of a fairly massive star collapses. The law of conservation of angular momentum magnifies any small original rotational velocity to an extremely high value, in the same way that a skater increases his speed of rotation by drawing in his arms. The magnetic field lines behave as if frozen into the stellar matter and co-rotate with it as it moves. When the star collapses, the lines are squashed together and the magnetic field is amplified (Figure 22).

In fact, in many ways a neutron star is an exaggerated version of a white dwarf. It has a radius of only about 15 kilometres: the reduction in size between a white dwarf and a neutron star is even greater than the reduction between the Sun and a white dwarf, and is comparable to that between a red giant and the Sun (Figure 23). The average density is not 1 tonne, but 100 million tonnes per cubic centimetre. The Sun turns on its axis once every 25 days,[3] whereas a neutron star completes a revolution in less than a second,

[3] In differential rotation, the rotational velocity is latitude-dependent.

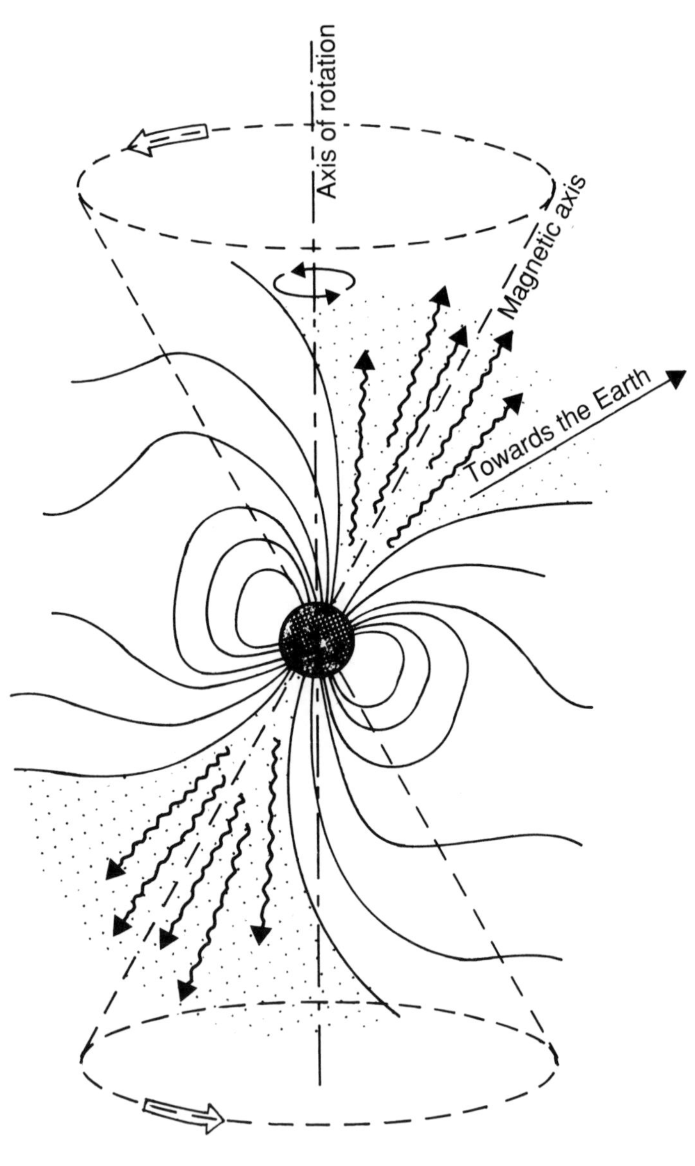

Figure 21. A pulsar model.

The neutron star's very strong magnetic field accelerates electrons which radiate radio waves along the magnetic axis. The latter does not coincide with the neutron star's rotational axis, so that on each revolution the radio beam may sweep the line of sight of a radio observer.

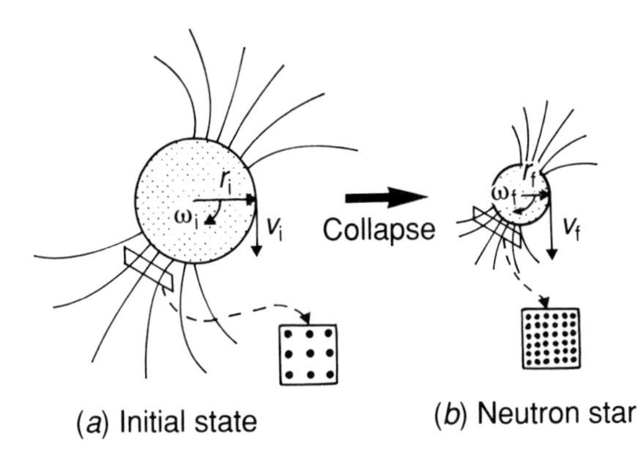

Figure 22: The formation of a pulsar.

This diagram illustrates why the rotational velocity and the intensity of the magnetic field increase during the formation of a neutron star by gravitational collapse. Before the collapse, the core of the pre-supernova of mass m, radius r_i and angular velocity w_i, has an angular momentum equal to $mr_i^2w_i$. After the collapse, the neutron star has approximately conserved its mass and angular momentum. As its radius r_f is much smaller than r_i, we deduce that the angular velocity w_f has increased in inverse proportion to the square of the radius. In the same way, if we suppose that the magnetic field remains attached to the contracting matter, its intensity, measured by the number of field lines crossing unit surface area, also increases by $1/r^2$.

in rigid rotation.[4] It is the same for the magnetic field: the Sun's magnetic field is similar to the Earth's, i.e. about 1 gauss, that of a white dwarf can reach 100 million gauss, whilst for a neutron star it is concentrated on a surface several billion times smaller and can reach up to a trillion gauss.[5] It was precisely these extreme properties that led to the detection of neutron stars.

Neutron stars cannot by seen in the optical part of the spectrum because their thermal luminosity, although emitted by a surface heated to 10 million K, is extremely low because of the small surface area; a body with a diameter of only 30 kilometres cannot

[4] It is believed that isolated white dwarfs spin either very slowly or not at all.

[5] The highest magnetic fields obtainable artificially in the laboratory are 300 000 gauss. They are produced by giant electromagnets weighing more than 10 tonnes.

be seen at distances of more than a few light years, which is much less than average interstellar distances. It was nevertheless found that there were optical counterparts of pulsars, including the Crab and Vela; the optical pulses were perfectly synchronised with the radio pulses. The Vela pulsar is thus one of the faintest stars detected in the sky, being 20 billion times less brilliant than Sirius.

Thus, the rotation and magnetic field produce periodic emission which can be detected not only at radio frequencies but also at higher frequencies. Even at X- and gamma-ray frequencies all the signals are modulated in the same way, by the star's rotation.

Cries and murmurs

It is thought that in some pulsars high energy radiation is emitted by the neutron star's polar caps, strongly heated by the impact of the charged particles channelled along the magnetic field and striking the ultra-hard crust at velocities close to that of light. A neutron star is simply a giant spinning magnet, which acts like a dynamo. At one revolution per second, a neutron star could create a potential difference of 10^{16} volts; under such conditions, the electrical forces can overcome the enormous surface gravity and release charged particles which are then accelerated. These particles immediately produce energetic gamma rays, but this radiation, caught by the magnetic field, finds it difficult to escape and is transformed into electron–positron[6] pairs. These pairs annihilate themselves in turn and create new gamma rays, which a short while later produce new electron–positron pairs, and so on until the radiation escapes from this zone. This process of multiple particle production is called a *cascade*, and enables several thousand particles to be created from a single particle released from the surface.

In the electromagnetic tornado produced by a pulsar, the radio emissions represent only a whisper; but it is this which is picked up by our instruments. Pulsar theoreticians work on models of the pulsar atmosphere (called magnetospheres because of the fundamental importance of the magnetic field) in an attempt to explain

[6] The positron is the electron's antiparticle.

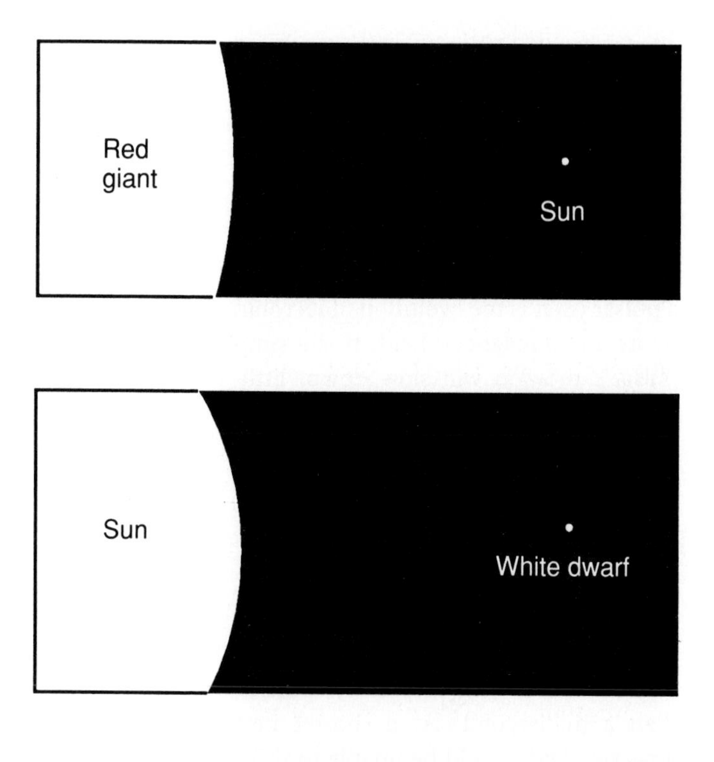

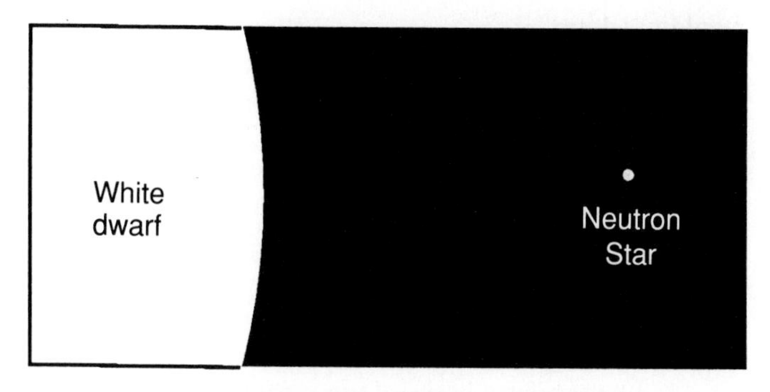

Figure 23: The comparative size of stars.

On this diagram, all the stars are assumed to have the same mass as the Sun. The reduction from the red giant to the Sun is 250:1, from the Sun to a white dwarf it is 100:1, from a white dwarf to a neutron star it is 500:1. A black hole is not shown because the star would have to be over 3 $M_{\odot}$ initially before it could become a black hole. Its final size would be similar to that of a neutron star.

all the details of a pulsar's emissions. This is similar to trying to understand the workings of a machine hidden in a factory, by listening to the muted sounds emanating from it.

The extinction of a pulsar

Just as the destiny of a star is controlled by its mass, that of a pulsar (that is the evolution of its rotational period) is determined by its initial magnetic field. It is a simple matter to predict that a pulsar's rotation will slow down, little by little, as its energy is dissipated. Since the magnetic field is responsible for the energy which is liberated, measurements of the rate at which the pulsar slows down can be used to calculate the neutron star's magnetic field.

For these reasons younger neutron stars usually rotate much faster than older ones. The Crab pulsar, very young as it was born in 1054, is naturally one of the most rapid: it completes 33 revolutions in each second, while the periods of the older pulsars may be several seconds long. However, a pulsar cannot have a period of less than a millisecond. At a shorter period than this, the pulsar, however rigid, would be unable to resist the centrifugal forces and would break up.

Pulsars slow down at rates in the range 10^{-12} to 10^{-19} second per second. These extraordinarily low values are nevertheless measurable over a period of several years. Once the rotation becomes too slow the pulsed emission ceases. The lifetime of a pulsar never exceeds more than a few million years.

Supernovae and pulsars

I have referred several times to the pulsars in the Crab and Vela nebulae which are associated with the remnants of famous supernovae. However, no pulsar has been detected in the other well-known remnants, such as Cassiopeia A, the Cygnus Loop, Tycho Brahe's supernova (1572) or Kepler's (1604). The association between pulsars and supernova remnants is itself remarkable: of the 450 known pulsars and 200 known supernovae remnants

(1991), only 3 (including the Crab and Vela) have been successfully paired.

A number of circumstances may have led to this surprising result. The simplest explanation is that after all a supernova does not leave behind a neutron star, but rather a different sort of stellar residue (complete disintegration or black hole), or that a neutron star is formed in the explosion but is displaced by the same explosion. In fact, the gravitational collapse of a parent star is probably not exactly spherical. As the rotational axis is generally not aligned with the magnetic axis, matter is ejected asymmetrically; at these velocities of over 10 000 km/s the ejection from a single side of the star of more than 10% of the total mass would give the pulsar a velocity of several hundred kilometres per second in the opposite direction. This phenomenon, like the recoil of a gun, results from the application of the law of conservation of momentum. The effect of the recoil may be to separate the supernova from the newly formed neutron star, forcing astronomers to look elsewhere for their pulsars.

It is also possible that many pulsars, like other stars, belong to binary systems; the explosion as a supernova of the massive companion to a pulsar could be violent enough to tear apart the binary and give the neutron star a velocity of the same order as actually observed (between 10 and 500 km/s).

Another plausible explanation for the almost total absence of pulsars in supernova remnants is that, although a neutron star is present, the pulsed emission phenomenon is either not active or is not observable from the Earth. An essential characteristic of pulsar emission is its anisotropy: a pulsar is a lighthouse, whose light is focused in a narrow cone inclined at an angle to the rotational axis. If the axis of the emission cone is unfavourably oriented, its beam will never sweep the Earth. Thus the neutron star is actually a pulsar but cannot be observed as such by terrestrial astronomers.

Pulsars are in general older than supernova remnants. The phase of pulsed radio emission occupies only a limited period during the lifetime of a neutron star: however, it is much longer than the lifetime of a supernova remnant. The average lifetime of pulsars can be estimated as about 3 million years from the rotational slowing down,[7] during which time supernova radio emission nebulae disperse completely. Finally, many more pulsars than supernova remnants have been observed. In our Galaxy alone the total number of pulsars could be as high as several tens of thousands.

Spinning tops in the skies

An 'ultra-rapid' pulsar spinning at 660 revolutions per second[8] was discovered in 1982. It is slowing down at such a microscopic rate that it is a more accurate clock than the best caesium clocks used on Earth to 'standardise' time.[9]

This ultra-rapid pulsar, called PSR 1937 + 21,[10] raises a particularly interesting theoretical problem. If its magnetic braking is so weak, its magnetic field must be 10 000 times less intense than those of the rapid pulsars in the Crab and Vela nebulae. But according to the traditional view of the formation of pulsars the weakness of the magnetic field implies great age, which is completely incompatible with its very high rotational velocity. How can these contradictions be reconciled?

A very attractive theoretical model suggests that the pulsar forms part of a *binary* system and that its rotation has been accelerated by a flow of gas drawn from its partner. This idea was confirmed by the recent discovery of two other ultra-rapid pulsars, one of 5.5 milliseconds and one of 6 milliseconds, which have very conspicuous companions. On the other hand, no partner has been found for PSR 1937 + 21; it is possible though that the pulsar had a very close white dwarf companion. The gravitational radiation emitted by this system would lead to a decay of the orbit until the two stars collided and the white dwarf disappeared, ripped apart by gigantic tidal forces. Its collision with the neutron star would have increased the pulsar's rotational velocity to the currently observed value.

[7] The oldest ones are however more than a billion years old.

[8] Its period is therefore 1.5 milliseconds.

[9] The slowdown rate of an ultra-rapid pulsar is 10^{-19} second per second. In other words, in one century its period decreases by only a billionth of a second.

[10] The numbers indicate the equatorial coordinates of the star: right ascension 19 h 37 min, declination + 21°.

Thus in the same way that the evolution of an ordinary star is modified if it belongs to a binary system, because of mass transfer from one star to another, the evolution of a pulsar in a binary system is different from that of an isolated one. Observation of certain pulsars having particular values of magnetic field and rotational velocity have shed new light on the formation of neutron stars. Some of them may belong to binary systems and may not have formed directly by the gravitational collapse of the core of a supernova, but by the continual growth of a white dwarf which, by capturing the gas of a nearby companion, eventually exceeded the Chandrasekhar limit and condensed into a neutron star, rather as the straw broke the camel's back.

Double pulsars to the rescue

At present, a dozen binary radio pulsars are known. One of their many attractions is that we are able to weigh neutron stars. Amongst them, PSR 1913 + 16, discovered in 1974, is far and away the most interesting. At 17 000 light years away in Aquila, PSR 1913 + 16 is truly a double pulsar since it is composed of a 1.4 $M_\odot$ neutron star which, when it was discovered, was pulsing at a rate of 16.94 radio pulses per second,[11] and a 'silent' companion of the same mass, *which is also a neutron star.* The two compact stars revolve around each other in 7 hours 45 minutes in an extremely close orbit of several million kilometres. Under these conditions the binary pulsar PSR 1913 + 16 offers an ideal test of the theory of General Relativity, which predicts that an accelerated mass should radiate energy as *gravitational waves.*[12] Dissipation of the orbital energy of a binary pulsar should lead to a contraction in the orbit, which manifests itself as a slow decrease over time of the orbital period.

Calculations based on Einstein's theory are in excellent agreement with observations carefully recorded over a period of 12 years. Most of the other theories of gravitation do not agree with these observations. The orbital period of PSR 1913 + 16 decreases each

[11] Its rotational velocity has since decreased.
[12] See Chapter 18.

year by 76 milliseconds. In about 300 million years, the two neutron stars will merge together emitting a final burst of gravitational radiation.

Starquakes

There is another phenomenon which alters pulsar rotation, but this time it is a sudden event which acts to accelerate the star. *Glitches*[13] reduce the period of the pulsar (i.e. increase its rotational velocity) by a ten millionth of a second over a period of a few days (Figure 24). Thus in February 1969 the Vela pulsar was suddenly spun up. Two other glitches occurred in 1971 and 1976. Several other pulsars have experienced glitches, including the Crab pulsar. The sudden increase in the rotational velocity is, however, quite small, so that after about a month of natural slowing caused by magnetic braking the neutron star has returned once more to the rotational velocity it had before the glitch.

Glitches can be explained as gigantic '*starquakes*', caused by instabilities which affect the crust of the neutron star and sharply modify its moment of inertia. A rapidly rotating neutron star is slightly flattened at the poles and broadened at the equator. From time to time the tensions in the surface caused by the flattening become too much and cracks appear, triggering a brutal readjustment, of the order of a millimetre only. None the less, a colossal amount of energy is brought into play: a quake on a neutron star would reach 25 on the Richter scale,[14] in comparison with the most violent quakes recorded on the Earth which have never exceeded 8.9.

For some astrophysicists, the fact that Vela has experienced several quakes has cast doubt on the validity of the model of surface quakes. The interval between two successive quakes as predicted by this model should be counted in centuries rather than years. Other explanations for glitches have been proposed, involving more

[13] This word is borrowed from electronics, where it refers to a sudden short-lived and unexpected event affecting the operation of an otherwise perfectly functioning piece of equipment.

[14] The Richter scale is used to measure the energy liberated by an earthquake. An increase of 1 on the Richter scale means an increase in energy of 20 times.

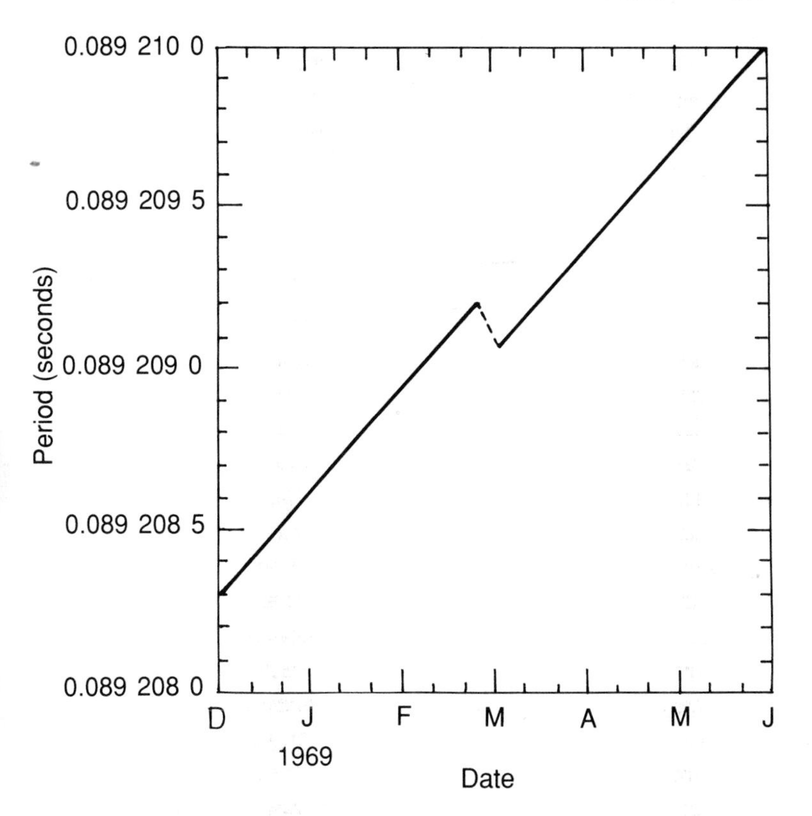

Figure 24: The glitch of the Vela pulsar.

A pulsar's period increases over time as the rotational velocity decreases. This decrease in the Vela pulsar's rotational velocity was suddenly interrupted between 24 February and 3 March 1969 by a glitch which accelerated it. During the glitch, the period decreased by only 200 billionths of a second. After the glitch the slowing down continued as before.

radical modifications of the neutron star's structure: turbulent motion in the deep layers, or even 'phase transitions' (similar to the passage from a liquid to a solid state) inside the nucleus, for which the envelope has to make adjustments.

Glitches also provide important information about the detailed internal structure of the neutron star. Here is a superb example of the help that astronomical observation can give to particle physics.

But how much do we actually know of the internal structure of neutron stars?

The interior of a neutron star

At first sight a neutron star is a giant atomic nucleus. The difference is that a neutron star is held together by gravity whereas an atomic nucleus is held together by nuclear forces.

In a neutron star, which is never more than a few kilometres across, gravity is such a strong force that it fixes the matter in very precise structures. One of its main effects is to erase all surface irregularities; the highest mountains on a neutron star are only a few centimetres high. The phenomena responsible for the electromagnetic emission of pulsars are all found within a minuscule layer heated to 10 million K.

The internal structure of a neutron star is still hypothetical, but a plausible description is as follows (Figure 25). The star is covered by a 1 kilometre thick shell of iron, a solid crystal of iron nuclei bathed in a sea of degenerate electrons. The density varies from 1 tonne/cm^3 (barely that of a white dwarf) to 400 000 tonnes/cm^3.

Below this is the 'mantle'. As we go deeper, the nuclei contain more and more neutrons but at the same time are less and less able to retain them; they become decayed in some sense. At a depth of about 5 kilometres the neutrons 'escape' from the nuclei and dissolve in the degenerate sea in which clumps of protons are floating. The density increases to 100 million tonnes/cm^3.

At a depth of about 10 kilometres, neutron matter constitutes the most important part of the star. The incredible pressures cause the crystalline structures to dissolve into a liquid which consists mainly of neutrons, protons and electrons. This liquid is probably a *superfluid*, a sort of perfect liquid, with strange properties: it has no viscosity. Viscosity tends to make any irregularities in a liquid disappear: for this reason, honey is more viscous than water. In a superfluid a vortex can maintain itself for several months.[15]

Finally, the composition of the solid core, about 1 kilometre in

[15] A superfluid can be produced in the laboratory by cooling helium to a temperature very close to absolute zero.

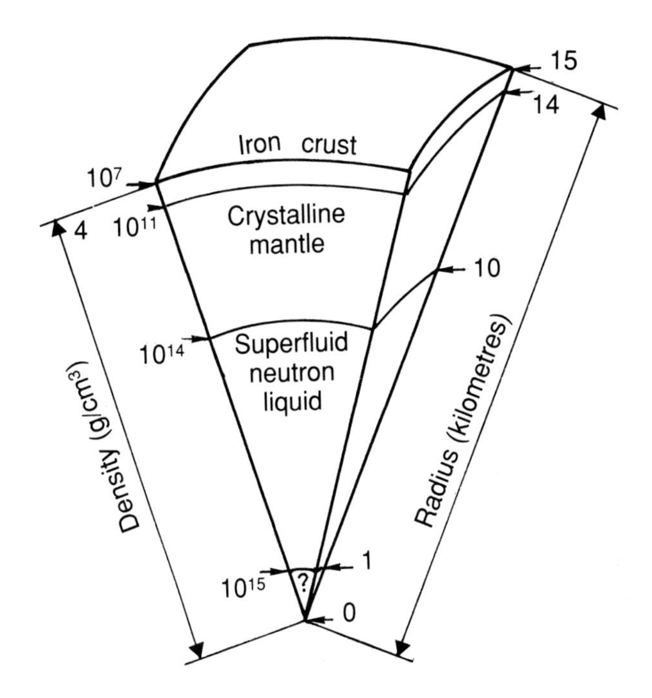

Figure 25: The internal structure of a neutron star.

radius, is very uncertain because at densities greater than a billion tonnes/cm³, we know very little about the possible states of matter that could exist. However we can speculate as to the nature of the elementary particles found in the nucleus, and various models with strange names have been devised: solid neutron crystals, pion condensates, quark matter, hadronic soup and so on.

The mysteries of dense matter

Neutron stars illuminate many fields in modern physics such as nuclear, atomic and plasma physics, relativity and electrodynamics, under extreme conditions of temperature, density, pressure and magnetic fields which cannot be reproduced in the laboratory.

We have clearly seen that to describe the interior of a neutron star we have to extrapolate from experimental physics which has not penetrated the secrets of matter at such high densities. Little is known about the *equation of state* of dense matter.[16] However, it can be confined between two limits: at one extreme a free gas, in which the particles are not subject to any forces, and at the other extreme a 'hard' state corresponding to maximum rigidity of matter, in which the velocity of sound equals that of light.[17]

All permitted states and all forms of matter lie between these two extreme states, but as far as neutron stars are concerned the choice between the many possibilities associated with these two states depends on the poorly understood physics of strong nuclear interactions between elementary particles

Fortunately, one very important property does not depend too much on the details of the equation of state of dense matter. This is the *maximum mass of a neutron star*. The reader will recall that white dwarfs are incapable of supporting a mass greater than 1.4 $M_\odot$, because beyond this limit the degenerate electrons which are the main component become relativistic and collapse under their own weight. For the same reason, neutron stars are unable to support an arbitrarily large accumulation of matter. The stability limit corresponds to the moment when the degenerate neutrons, under enormous gravitational pressure, become relativistic themselves.

To calculate the maximum mass of neutron stars with the same accuracy as that of white dwarfs we would need to know the equation of state of degenerate neutrons with the same accuracy as that of degenerate electrons, which we do not. However, a good approximation to the limiting mass can be obtained by the following reasoning. The density of a neutron star increases from the crust to the centre; at an intermediate point it reaches nuclear density and from this point on the equation of state can be used only with caution. Therefore the equation of state at sub-nuclear densities – known experimentally – can be used to describe the envelope around the neutron star. The core is described by the equation for the most rigid state possible. The two solutions are

[16] This is the law governing the variation of certain thermodynamic quantities such as pressure, as a function of other quantities, such as density.

[17] The velocity of sound in a material increases with its rigidity; it is 330 m/s in air, 1500 m/s in water and 5 km/s in steel.

added and the total mass is the mass of the envelope plus the mass of the core.

The limit obtained in this way is 3.2 $M_\odot$. Even this limit is an overestimate and more sophisticated models give a value between 2 and 3 $M_\odot$. This result is fundamental, because it immediately raises a question: what is the result of the gravitational collapse of very massive stars?

8

Gravitation triumphant

'What would physics look like without gravitation?'

A. Einstein (1950)

White and black dwarfs, neutron stars and pulsars are stellar corpses which are not too disturbing. But black holes? Michell and Laplace imagined that large invisible stars might exist, but they had no idea of the mechanism which might form them and they did not envisage black holes with a solar mass. They lacked knowledge of quantum mechanics and General Relativity, theories which would not be developed until later.

The reappearance of black holes as a possible result of gravitational collapse occurred in 1939, when the American physicists Robert Oppenheimer (who had already worked on the theory of neutron stars) and Hartland Snyder were studying the collapse of a simplified 'model star', which was spherical and without internal pressure, using the equations of General Relativity. They discovered that in certain circumstances the gravitational pressure was so great that a stable neutron star could not form; on the contrary, nothing was able to stop the star contracting until it reached a 'point' of zero volume and infinite density. Long before this stage was reached, the contracting star ceased to have any communication with the outside Universe.

The theoretical prediction of the existence of stellar black holes thus rests on three key points:

1 There is no force in Nature which can support more than 3 $M_\odot$ of 'cold' matter, that is, matter which is inert to thermonuclear reactions.
2 Many observed hot stars have masses much greater than 3 $M_\odot$.

3 The timescale required for a massive star to consume its nuclear fuel and be subjected to gravitational collapse is a few million years, and such events have already occurred in our Galaxy which is over 10 billion years old.

The weak point of this argument is the assumption that a massive star is capable of developing a *degenerate core* – the only thing that collapses – which exceeds the stability limit for neutron stars. The largest known stars have masses reaching 100 $M_\odot$.[1] However, during their evolution, all stars lose some of their mass in a 'stellar wind'. In the case of the Sun and other modest sized stars, this loss is minimal on the Main Sequence; mass ejection occurs mainly at the end of nuclear evolution in the planetary nebula phase. However, there are good reasons to think that very large stars eject large quantities of matter from the beginning of their existence. Our knowledge of this problem, theoretical and observational, is so slight that it is not possible to draw firm conclusions, nor even to dismiss the extreme hypothesis according to which, whatever the initial mass of a star, its mass loss by stellar wind will always be sufficient to reduce the mass below 3 $M_\odot$. In this event, the formation of black holes in supernovae would be quite simply impossible.

However, as will be seen in Part 4, we believe that black holes of several $M_\odot$ have actually been detected in certain X-ray sources. As our knowledge stands, it is more reasonable to think that all supernova precursors between 10 and 100 $M_\odot$ will lead to either a neutron star or a black hole. Detailed models of the explosion of a supernova, calculated with powerful computers, show that there are two possible cases in which a black hole will form.

1 When the mass of the degenerate core is greater than the stability limit of neutron stars the collapse will lead directly to a black hole; but it is not known if its formation involves an ejection of matter (the stellar envelope will not be able to rebound off a hard nucleus as in the case of a neutron star).
2 When the mass of the nucleus is less than the critical value but little mass is ejected, a neutron star would be created

[1] The record is presently held by the star HD 698, which has a mass of 113 $M_\odot$.

at first; but it would be unable to support the weight of the envelope and would collapse into a black hole.

Apart from these two possibilities for the creation of black holes of several $M_{\odot}$ in supernovae, there is a third possibility, which would take place in stages over a long period of time. This involves the prior formation of a neutron star in a supernova followed by a long phase of capture and deposition of matter onto the neutron star's surface (the most favourable situation obviously being a binary system) until the total mass violates the critical stability limit. Such a mechanism is similar to the transition of a white dwarf into a neutron star, and for it to work the deposited gas should not be regularly expelled by surface nuclear reactions as in novae.

In conclusion, a black hole appears in the scenario of stellar evolution as the final triumph of gravity in regulating the life of a star (see Appendix 2). But there is more to it than this. Gravitation governs all the large assemblages of matter in the Universe. We shall see later that a dense cluster of stars also evolves by contracting its core and could develop black holes not of several $M_{\odot}$, but of a thousand, a million or even a billion $M_{\odot}$. We will also see that a black hole can grow by sucking in matter and that from a dwarf it can turn into a giant, into one of the invisible stars that Michell and Laplace imagined. And then there are also minuscule black holes much too light to have collapsed under their own weight, but 'forced' into existence by a gigantic external pressure that only the early Universe could have produced.

The black hole is a strange sort of corpse. Once formed, it is not 'dead' but destined to have a life 'full of sound and fury'. The second half of this book will recount its story.

PART 3
LIGHT ASSASSINATED

'If one intends to abandon Relativity, here is the place to do so. Otherwise one is on the way to a new world of physics, both classical and quantum. Here we go!'

B. Harrison, K. Thorne, M. Wakano and J. Wheeler (1965)

9

The far horizon

'In the deathless boredom of the sidereal calm, we cry with regret for a lost Sun . . .'

Jean de La Ville de Mirmont, *L'Horizon chimérique*

The Schwarzschild solution

In December 1915, a month after Einstein published his equations of General Relativity, the German physicist Karl Schwarzschild discovered the solution which described the gravitational field surrounding a sphere in a vacuum. From the Russian front where he was fighting,[1] Schwarzschild sent his manuscript to Einstein and asked him to take care of the publication details. Einstein was very impressed. He replied, 'I had not expected that the exact solution to the problem could be formulated. Your analytical treatment of the problem appears to me splendid.'

There are two reasons why Schwarzschild's space-time geometry is so interesting. In the first place, it is a very good description of the gravitational field in the Solar System. The Sun is practically spherical, and the mass of the surrounding matter is so low that one can regard it as a vacuum. The trajectories of all the light rays and free-falling bodies, such as the planets and comets, therefore follow the geodesics[2] of Schwarzschild's curved space-time. They can be calculated with precision and agree with the observed values of light rays passing close to the Sun and the advance in the perihelion of the planets, which Newton's gravity is unable to explain.

On the other hand, Schwarzschild's solution is of universal

[1] A patriot, Schwarzschild had voluntarily enlisted in the Prussian army. At the time he discovered his solution, he contracted an incurable disease, pemphigus. He was quickly repatriated and died in May 1916.

[2] The equivalent of straight lines in a curved geometry, see page 41.

interest because it is independent of the type of star which produced it. It depends on only one parameter, the mass. The gravitational field around the Sun and that of a neutron star is identical provided the masses are the same. A point mass would also produce the same result.

This is where the difficulties begin. The behaviour of the geometry becomes disconcerting as it approaches the point source of gravitation. More precisely, it begins to behave strangely at the critical distance given by $r = 2GM/c^2$, where M is the mass of the central star, G is Newton's constant of universal gravitation and c is the velocity of light.[3] The distance is proportional to the gravitational mass. For the mass of the Sun it is 3 kilometres, it is 3 million kilometres for a million solar masses and 1 centimetre for the Earth. The *Schwarzschild radius* is nothing other than the critical size of a star below which the escape velocity from its surface, calculated '*à la* Newton', reaches the velocity of light. Without knowing it, Schwarzschild had just reopened the door on Michell and Laplace's forgotten speculations about invisible stars!

The magic circle

There seemed to be two pitfalls on the route from the Schwarzschild solution to a theory of black holes: one was mathematical and the other astronomical.

In Schwarzschild's version, space and time lose their identity inside the critical radius $r = 2M$. The 'rules' used to measure distances and time outside the radius become absurd; one runs off to infinity and the other becomes zero. Eddington described this 'singularity' in the geometry in these terms; 'There is a magic circle which no measurement can bring us inside.'

The question of the magic circle led to a heated discussion at the Paris Colloquium in 1922. The best group of relativists that could be imagined gathered around Einstein, including Jean Becquerel, Henri Brillouin, Elie Cartan, Jacques Hadamard and Paul

[3] In the following I shall take the simple version $r = 2M$ of this formula. This amounts to choosing the units of mass, length and time so that G and c both have the value 1.

Langevin. However, this array of theoretical physicists was unable to solve the mathematical problem posed by the critical radius. At the very most they sensed the possibility of gravitational collapse.

For a long time the magic circle was considered a flaw in the theory of General Relativity and hindered its development. It was only in the 1950s that theorists agreed on the interpretation of the singularity at the Schwarzschild radius. The 'pathological' behaviour of the geometry is just a mathematical accident. David Finkelstein proved that it was the result of a bad choice of coordinates.[4] Many years earlier Eddington had discovered a coordinate system in which Schwarzschild geometry lost the magic circle. But he was unable or unwilling to see the consequences, because he was preoccupied with another problem in astronomy, that of gravitationally condensed stars.

The reappearance of invisible stars

The idea that a star like the Sun could condense itself into a sphere with a 3 kilometre radius was as unacceptable at the beginning of the twentieth century as it was in Laplace's day because it required a density of matter which defied the imagination. In 1931 the Japanese physicist Yusuke Hagihara produced an impressive piece of mathematics, in which he calculated all the geodesics in Schwarzschild's space-time, including those which penetrated the 'magic circle'. He concluded, 'In fact, it is quite improbable that in any star the distance $r = 2M$ from the centre lies outside its radius. In order that the radius of a star with mass comparable with our Sun be equal to the distance $r = 2M$, its density ought to be about 10^{17} times that of water, while in the densest star, the companion of Sirius, a white dwarf, the density is about 6×10^4 times that of water.[5] There is no such diversity in the masses of stars as to overcome this tremendous high magnitude of the critical density. Therefore the orbit inside $r = 2M$ is physically highly improbable.'

[4] In General Relativity, all coordinate systems are equivalent for describing physical phenomena, but for some of them the calculations are much simpler than for others.

[5] Later observations showed that white dwarfs were 10 times denser than Hagihara calculated, see Chapter 5.

This quotation sums up perfectly the pragmatic attitude held by most astrophysicists. They were only interested in the outer regions of the Schwarzschild geometry as it could be applied to the Solar System, and they completely ignored the strange behaviour at the critical radius.

Some of them dared to go further, however. In 1920, A. Anderson asked himself what would happen to a star which actually shrank to a volume close to its 'magic circle'. 'Thus if the body of the Sun should go on contracting there will come a time when it is shrouded in darkness, not because it has no more light to emit, but because its gravitational field will become impermeable to light.' A year later Sir Oliver Lodge repeated almost word for word Michell and Laplace's reasoning: 'if light is subject to gravity . . . a sufficiently massive and concentrated body would be able to retain light and prevent its escaping . . . If a mass like that of the Sun could be concentrated into a globe of about 3 kilometres in radius, such a globe would have the properties above referred to: but a concentration to that extent is beyond the range of rational attention . . . But a stellar system – say a super spiral nebula – of aggregate mass equal to 10^{15} suns might have a group radius of 300 parsecs with a corresponding average density of 10^{-15} g/cm^3 without much light being able to escape. This really does not seem an utterly impossible concentration of matter.'

In the final analysis, if astrophysicists still have trouble accepting the fabulous densities implied by the stars of several $M_\odot$ condensed to less than the Schwarzschild radius, then some of them are ready to accept the possibility of the existence of much greater masses simply because the corresponding densities become 'reasonable', that is, comparable to those already observed in nature.

At the same time, the new theory of quantum mechanics supported the hypothesis of gravitational collapse by predicting the existence of degenerate states, which were much denser than anyone had ever dared to imagine. The world was ready for the reappearance of the idea of invisible stars. But the time had not yet come. Eddington was paradoxically the greatest defender of General Relativity and the fiercest opponent of the idea of a star condensed within the Schwarzschild radius. 'I think there should be a law of Nature to prevent a star from behaving in this absurd way!' In support of his beliefs, Eddington had to modify Fermi's law of degeneracy to allow any cold masses, whatever their size, to remain in equilibrium. He developed his ideas at the International Astronomical Union in 1935, of which he became president three years later. The young Chandrasekhar slipped a note to the president of the session asking for permission to put an opposing view. He was refused; Eddington was so famous that his opinion could not be doubted!

In spite of these events the march of history could not be halted. Chandrasekhar became famous by developing the first models of condensed stars: white dwarfs. The theory of gravitational collapse was truly born in 1939, thanks to the work by Oppenheimer and Snyder (see Chapter 8). They used the equations of General Relativity to calculate the gravitational collapse of a spherical mass below the Schwarzschild radius. They were able to show rigorously that matter, and with it space-time itself, would collapse to form a region from which not even light could escape.

The term *black hole* was used for the first time on 29 December 1967, during a lecture given in New York by John Archibald Wheeler. The brilliant career of black holes could begin . . .

Darker than you think

The Indian astrophysicist Jayant Narlikar related the following anecdote. In the eighteenth century, in Calcutta, there was a fortress called Fort William. It contained a small dark room called the 'black hole of Calcutta'. This room measured five metres by four and was designed to take three prisoners. In 1757, there was a bloody revolt in Bengal. As a reprisal, the cruel governor of the province imprisoned 46 prisoners from the enemy army in the black hole of Calcutta. They spent 10 hours in the room, during the hottest part of the summer. Only 22 of them survived.

This story is so awful that some historians doubt its authenticity. Whatever the truth, it symbolises the idea of a gluttonous black hole devouring everything which passes through its door. This macabre feature of black holes has been widely publicised by certain sections of the media; however, it is only one of their many aspects. A black hole is an 'object' which is at once simple and extremely disconcerting in the way it distorts space-time. First of all, let us dissect the traditional image of a black hole, that of a *cosmic prison.*

We return to the fundamental description of a black hole: *a region of space-time inside which the gravitational field is so intense that it prevents all matter and radiation from escaping.* A strong gravitational field indicates a high concentration of matter. To 'build' a black hole, it is necessary to place a given mass inside a certain critical volume whose size is given, in the spherical case, by the Schwarzschild radius. Table 3 shows how different a black hole is from bodies such as atoms and stars.

Ignoring for the moment possible mechanisms for producing black holes, it is theoretically possible to have black holes of all sizes and masses. There could be microscopic black holes, the size of an elementary particle but with the mass of a mountain, or black holes of several $M_\odot$ and several kilometres in diameter, or giant black holes of several billion $M_\odot$, as big as the Solar System itself (see Appendix 2). Consequently, and contrary to popular opinion, the average density of black holes does not have to be very high. It is inversely proportional to the square of the mass. Of course, a 10 $M_\odot$ black hole resulting from the gravitational collapse of a star beyond the stage of a neutron star would have a 'nuclear' density of 10^{14} g/cm^3, but a black hole of several billion $M_\odot$ would have a density 100 times less than that of water. A black hole does not have to be an extremely dense star, it just has to be compact enough to imprison light.[6]

The imprisonment of light

'Here day fights with night.'

The last words of Victor Hugo

Let us suppose that a perfectly spherical star surrounded by vacuum collapses within its Schwarzschild radius. Its surface is

[6] The notion of a body's density differs from that of compactness, in that the density is the ratio of the mass to the volume, whereas the compactness is the ratio of the critical radius to the actual radius (see Table 3).

Table 3: *The gravitational parameter of ordinary bodies*

Object	Mass	Size R	Schwarzschild radius Rg	Gravitational parameter Rg/R
Atom	10^{-26} kg	10^{-8} cm	10^{-51} cm	10^{-43}
Human being	100 kg	1 m	10^{-23} cm	10^{-25}
Mountain	10^{12} kg	1 km	10^{-13} cm	10^{-18}
Earth	10^{25} kg	10^{4} km	1 cm	10^{-9}
Sun	10^{30} kg = 1 $M_\odot$	10^{6} km	1 km	10^{-6}
White dwarf	1 $M_\odot$	10^{4} km	1 km	10^{-4}
Neutron star	1 $M_\odot$	10 km	1 km	10^{-1}
Galaxy	10^{11} $M_\odot$	10^{5} l.y.	10^{-2} l.y.	10^{-7}
Universe (if closed)	10^{23} $M_\odot$	10^{10} l.y.	10^{10} l.y.	1 ?

Note: The gravitational parameter is the ratio between the Schwarzschild radius – which depends only on the mass – and its real size. In other words, it measures the 'compactness' of a body; the closer its parameter is to one, the closer a body is to the black hole state. The numerical values in the table are given to the nearest power of ten. The parameters for the Universe require careful consideration: see Chapter 19.

hot and emits radiation. How does light become progressively imprisoned and the star become a black hole?

Where Michell and Laplace referred to the escape velocity, General Relativity is more subtle. In 1923, G. Birkhoff showed that Schwarzschild's solution describes not only space-time around a static mass but also that around a collapsing or expanding star, so long as it remains exactly spherical. If the Sun started to vibrate, expanding and contracting at the same rate in all directions, or even if it was replaced by a black hole of the same mass, the geometry of the Solar System would not change. The orbits of the planets and comets would not be affected at all. There would simply be no more light. Birkhoff's theory indicates that the behaviour of light rays emitted by a spherically contracting star is faithfully described by the geodesics of Schwarzschild's geometry.

Figure 26 shows four stages of gravitational collapse of a spherical star which progressively retains more and more light. Before the collapse (Figure 26*a*), the mass of the star fills a volume which is much greater than that described by the Schwarzschild radius. According to the theory of General Relativity, its gravitational field has very little effect on the 'elastic tissue' of space-time. The light leaving a point on the surface of the star can escape from the surface in a straight line in any direction.

Then the star collapses (Figure 26*b*). As its radius approaches the critical Schwarzschild radius, the gravitational well deepens and the curvature of space-time increases. By virtue of the Principle of Equivalence, light rays are compelled to follow the curvature and so are deviated from a straight line in following the geodesics. When the star's radius equals 1.5 times the Schwarzschild radius, light rays emitted *tangentially* 'fall back' onto the surface like water from a fountain. They form a 'photon sphere', a sort of cocoon of light surrounding the collapsing star, from which distant astronomers would occasionally see a few escaping photons.

As the gravitational collapse continues fewer and fewer photons are able to escape. The light's 'escape cone' continues to shrink (Figure 26*c*). When the star reaches the critical Schwarzschild radius, all light rays are recaptured, even those which were emitted *radially* (that is perpendicular to the surface). The escape cone has completely closed and the photon sphere has disappeared (Figure 26*d*). A black hole has formed. Its surface, the Schwarzschild sphere, is a frontier beyond which it is impossible to see. It is an *horizon*.

The event horizon

The terrestrial horizon is caused by the curvature of the planet and is a frontier beyond which it is impossible to see. The Earth's horizon is relative. It is a circle centred on the observer, and moves with him.

By contrast, the horizon of a black hole is absolute. It is a frontier in *space-time*, independent of the observer, dividing events into two categories. Outside the horizon it is possible to communicate over arbitrary distances by light signals. This is the ordinary universe we inhabit. Within the horizon, light rays are not free to travel from one event to another but are focused towards the centre. Communication between events is subject to severe constraints. This is a black hole.

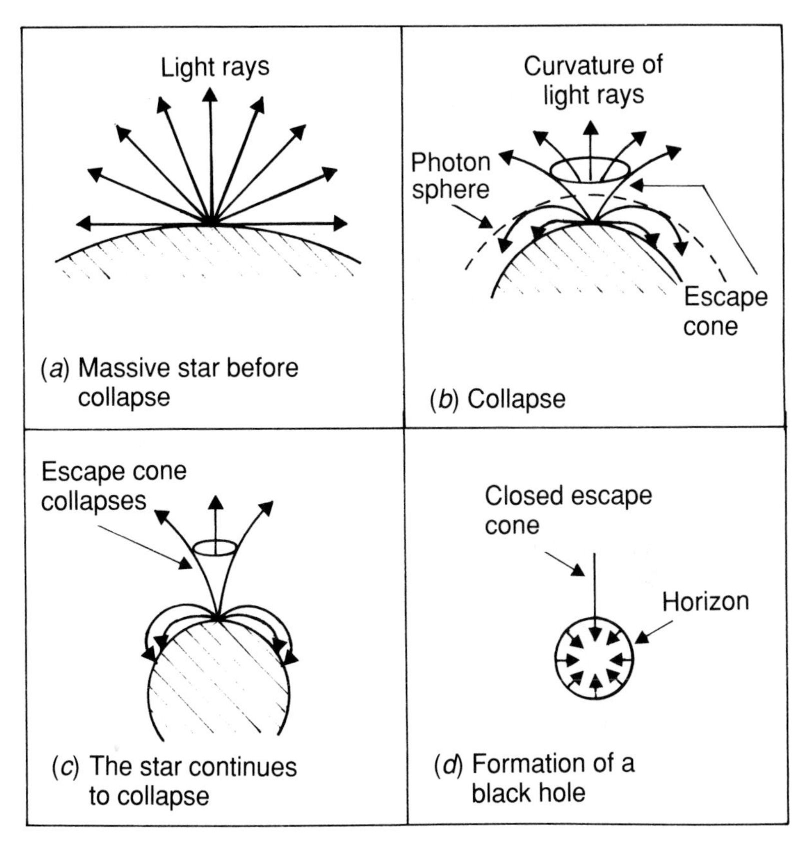

Figure 26: The imprisonment of light in four stages. (*a*) Massive star before collapse; (*b*) collapse; (*c*) the star continues to collapse; (*d*) formation of a black hole. (After W. Kaufmann.)

Figure 27 is a space-time diagram showing the Schwarzschild geometry around a spherically contracting star, resulting in the formation of a black hole. This is the most important diagram in the book because it provides the fundamental key to a proper understanding of black holes. For this reason it deserves special attention.

As with all space-time diagrams, the curvature is visualised by means of light cones. It should be remembered that at each event a

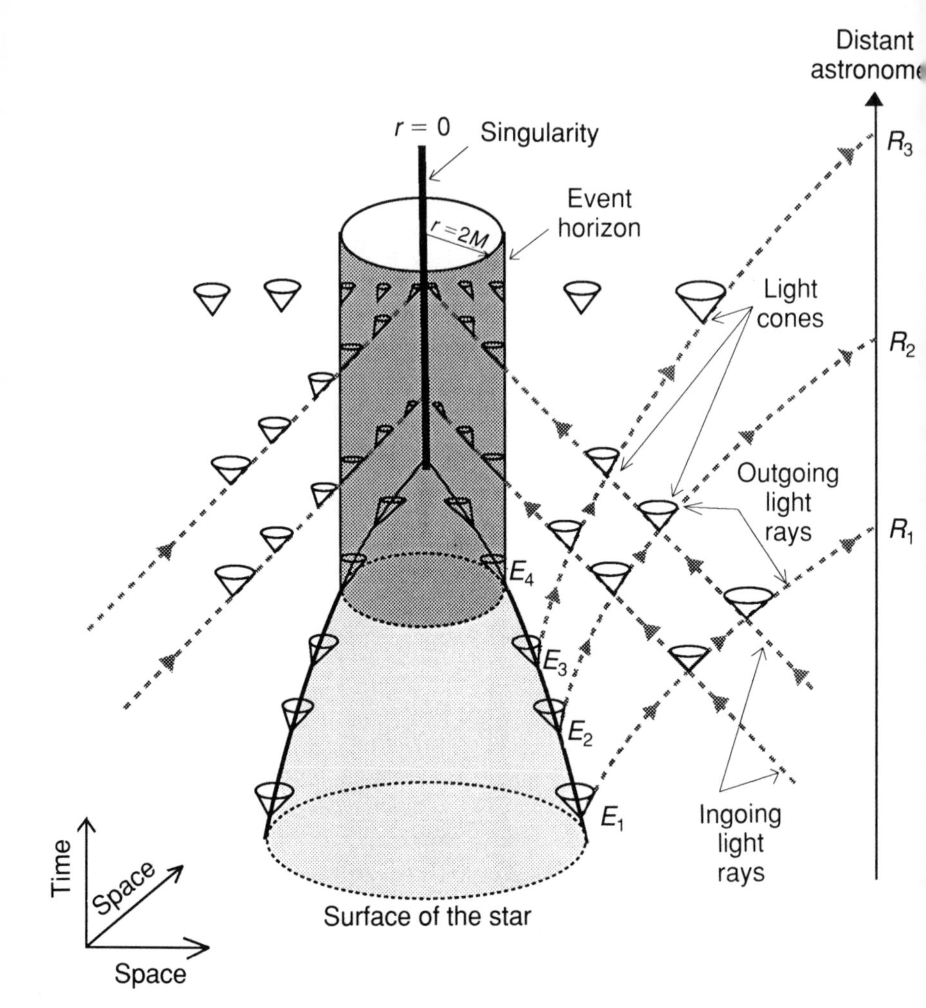

Figure 27. A space-time diagram showing the formation of a black hole by gravitational collapse.

The diagram shows the complete history of the collapse of a spherical star from its initial contraction to the formation of a black hole and a singularity. Two space dimensions are measured horizontally, and time is shown on the vertical axis, measured upwards. The centre of the star is at $r = 0$. The surface of the star at a given instant, normally a sphere, is reduced to a circle because one space dimension has been suppressed. The curvature of space-time is visualised by means of the light cones generated by the trajectories of light rays. Far

light cone is generated by the trajectories of electromagnetic waves and confines the worldlines of all particles incapable of travelling faster than the velocity of light. In the absence of gravity, all the light cones are 'parallel' to each other, in the sense that their generators are inclined at 45° and their opening angles are 90° with a suitable choice of the units of space and time, (see Figure 4.) They describe the flat space-time of Minkowski which is the basis of Special Relativity. In the presence of a gravitational field and the associated curved geometry the light cones are deviated and their openings are smaller.

In order to simplify the diagram, only the light rays propagating in a radial direction (entering and leaving) are shown. The photon sphere discussed earlier therefore does not appear. Far away from the collapsing region, space-time is almost flat and the light cones are straight. In fact the gravitational field caused by the central mass decreases with distance and the curvature in space-time lessens. We say that the Schwarzschild space-time is *asymptotically flat*, which means that at a great distance from the mass it becomes identical to Minkowski space-time.

Closer to the source of the gravitational field the curvature increases and influences the light cones, which close and incline towards the centre of the collapsing region. The light rays find it

Caption for Fig. 27 (*cont.*)

away from the central gravitational field the curvature is so weak that the light cones remain straight. Near the gravitational field the cones are deformed and deviated by the curvature. On the critical surface of radius $r = 2M$, the cones are tipped over at 45° and one of their generators becomes vertical so that the allowed directions of propagation of particles and electromagnetic waves are oriented towards the interior of this surface. This is the event horizon, the boundary of a black hole (grey region). Beyond this the stellar matter continues to collapse into a singularity of zero volume and infinite density at $r = 0$. Once a black hole has formed and after all the stellar matter has disappeared into the singularity, the geometry of space-time itself continues to collapse towards the singularity, as shown by the light cones. The emission of the light rays at E_1, E_2, E_3 and E_4 and their reception by a distant astronomer is discussed in the text on page 131.

more and more difficult to escape. There comes a point when the light cones have swung over 45° and one of their generators has become vertical, so that all the allowed directions of propagation are focused towards the centre of the gravitational field. This is the point at which light is imprisoned, and corresponds to the formation of the event horizon at $r = 2M$. Beyond this the light cones are even more inclined and their openings much narrower. The trajectories of all material particles confined within the light cones converge inexorably on the vertical line $r = 0$. This geometric centre of the black hole truly is a *singularity*, where all matter and space-time curvature are compressed indefinitely.

The formation of a black hole divides space-time into two parts separated by the event horizon. Matter and radiation are able to pass from the outside to the inside, but not in the other direction. This is why it is called a 'black hole'.

The imprudent traveller

At a great distance from a black hole, space-time is similar to that of the Solar System, slightly curved by the central mass of the Sun. However, whereas the Schwarzschild geometry stops at the surface of the Sun, 700 000 kilometres from its centre, in the case of the black hole it continues right up to the central singularity. Of course it is only near the horizon that the characteristic distortions associated with a black hole are manifested.

Like all gravitational sources, a black hole causes *tidal forces.*[7] For an astronaut falling head first towards a black hole, his feet are attracted less than his head. His body is stretched out by the tidal forces which increase as he approaches. Of course the human body cannot survive such stretching, nor pressures greater than 100 times atmospheric pressure (which is already 1 kg/cm^2). An astronaut attracted towards a 10 $M_\odot$ black hole – with a radius of 30 kilometres – would be killed by the tidal forces long before he reached the horizon, at an altitude of 400 kilometres. At the horizon the stretching effect would be the same if he was hanging

[7] This is the translation in Newtonian language of the curvature of space-time (see page 37).

from a girder of the Eiffel tower with the entire population of Paris suspended from his ankles!

However, the intensity of the tidal forces depends on the density of the matter which causes them. More massive black holes have lower densities and the exterior space-time is less curved. Consequently a human being could survive near a very massive black hole. Our guinea-pig astronaut could reach the horizon of a 1000 $M_\odot$ black hole and he would even be able to explore the interior of a giant 10 million $M_\odot$ black hole, since the tidal forces at the horizon would be weaker than those produced by the Earth, which are already imperceptible. However, once he had crossed the frontier the astronaut would be inexorably captured by the central singularity and then, whatever the mass of the black hole, he would be torn to pieces by the infinite tidal forces!

Frozen time

Figure 27 also shows how light rays produced at events E_1, E_2, E_3 and E_4 leave the surface of the contracting star and at R_1, R_2, R_3 and R_4 reach an astronomer situated at a great distance (whose worldline is represented by a vertical straight line). We assume that the time intervals between the four events are equal when they are measured by a clock placed at the surface of the star and participating in its collapse. The space-time diagram proves, however, that the reception intervals between R_1, R_2, R_3 and R_4 become longer and longer. At the limit, light rays emitted from E_4, just at the point where the event horizon is forming, take an infinite time to reach the distant astronomer. (The point R_4 is therefore not represented on the diagram.)

This phenomenon of 'frozen time' is an illustration of the extreme elasticity of time predicted by Einstein's relativity, according to which time runs differently for two observers with a relative acceleration.[8] The surface of a gravitationally collapsing star is effectively accelerating with respect to the distant astronomer who is not participating in the free fall. Consequently the *proper time* of the collapse, measured by a clock placed on the surface of the star,

[8] Or, from the Principle of Equivalence, in different gravitational fields.

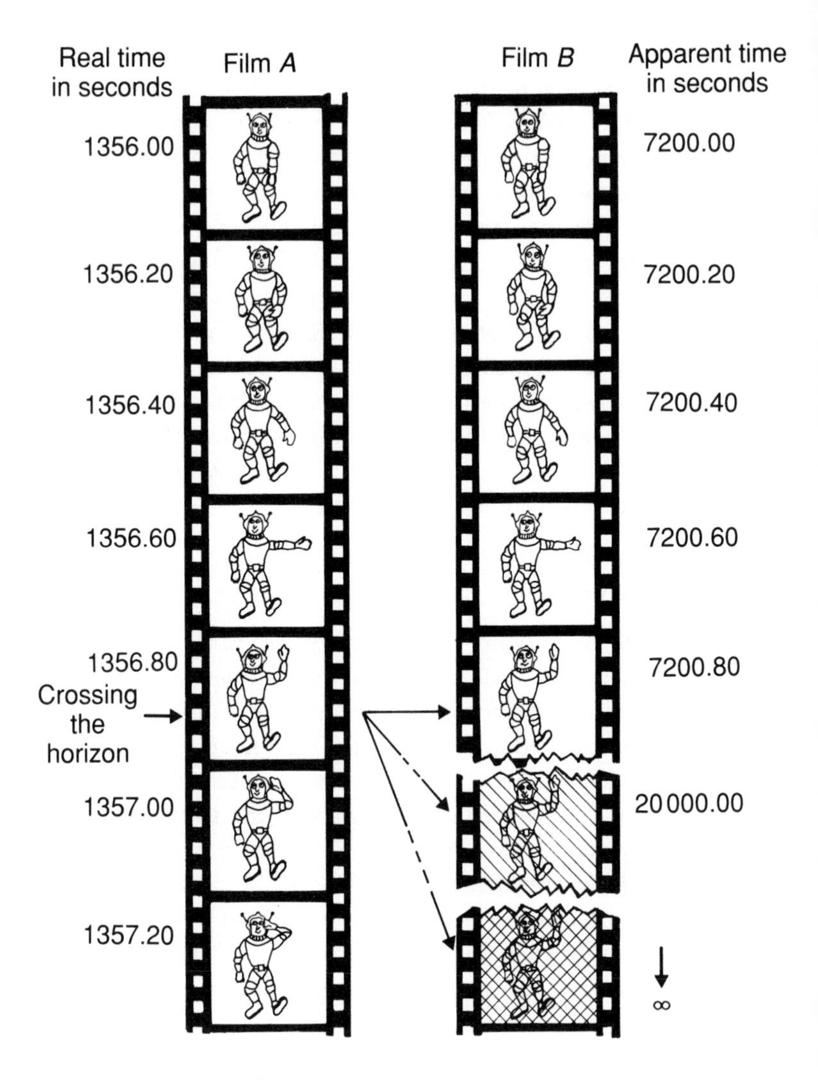

Figure 28. The astronaut's salute.

The film on the left shows the scene on board the spaceship in proper time, that is, as measured by the ship's clock as the ship falls into a black hole. The astronaut's salute is decomposed into instants at proper time intervals of 0.2 second. Crossing the horizon is not accompanied by any particular event.The film on the right shows the scene received by the distant observers via televised images. It is also decomposed into intervals of apparent time, also of 0.2 second. At the beginning of his

is very different from the *apparent time* of the collapse, measured by an independent and distant clock. The contraction of the star below the Schwarzschild radius happens in a finite proper time, but in an infinite apparent time. A distant astronomer will *never* be able to see the formation of a black hole, or ever see beyond it.

The freezing of apparent time revealed by the longer intervals of reception of the signals, also manifests itself in a decrease in the apparent frequency of radiation leaving the star, since the frequency is the number of oscillations of the light ray per second.[9] If the apparent frequency of the radiation decreases, then its wavelength appears to increase, and we say that it is redshifted, recalling that the longest wavelength of visible light corresponds to the colour red (see Table 1). The distant astronomer will see the collapse not just increasingly slowed down, but also more and more reddened and ever fainter.

Figure 28 is a more picturesque illustration of frozen time. A spaceship has the mission of exploring the interior of a black hole – preferably a big one, so that it is not destroyed too quickly by the tidal forces. On board the ship the commander sends a solemn salute to mankind, just at the moment when the ship crosses the horizon never to return. His gesture is transmitted to distant spectators via television.

Film *A* shows the series of images taken at equal intervals in the astronaut's *proper time*. It represents the scene as observed by the crew on board the ship. The commander's salute begins at 1356.00

Caption for Fig. 28 (*cont.*)

gesture the salute is slightly slower than the real salute, but initially the delay is too slight to be noticed, so the films are practically identical. It is only very close to the horizon that apparent time starts suddenly to freeze; the film on the right then shows the astronaut eternally frozen in the middle of his salute, imperceptibly reaching the limiting position where he crossed the horizon. Besides this effect, the shift in the frequencies in the gravitational field causes the images to weaken and they soon become invisible.

[9] This is Einstein's effect, already mentioned in Chapter 3.

seconds on the on-board clock and ends at 1357.20 seconds. The horizon crossing occurs during the gesture, and is not accompanied by any particular phenomenon. For the explorers on the ship, the frontier of the black hole has nothing 'magical' about it.

Film *B* shows the series of images received on the screens of distant spectators, at equal intervals of *apparent time*. Initially the film is identical to film *A*, but as the ship approaches the horizon it slows down. The distant observers continue to receive practically identical images, showing the astronaut permanently frozen in the position at which he crossed the frontier. In fact because of the shift in the frequencies and the decrease in the intensity, the apparent images soon become too weak to be visible. For the spectators all of the ship's voyage inside the black hole is thus lost. The television image transmitted by the ship just as it crossed the frontier escapes to infinity, and all the subsequent images transmitted by it cannot escape from the black hole, but fall into the singularity instead.

The frozen time measured by an observer – however close he is to the event horizon – is such a striking property that the term *frozen star* was often used to describe black holes.[10] It was finally abandoned because it described only a minor aspect of the physics of black holes. If the horizon of a black hole reaches to the infinite future of exterior spectators, the interior of the black hole is not fantasy for all that. The theory of General Relativity enables us to explore the interior of black holes (without worrying about tidal forces!).

The inverted world

'All hope abandon, ye who enter here.'

Dante, *Inferno*

Unlike the other compact stars such as white dwarfs and neutron stars, in which gravitational collapse has been interrupted by the internal resistance of matter, and which have hard surfaces, nothing can stop the collapse once the Schwarzschild radius has been passed and an event horizon has formed.

[10] Above all by Russian astrophysicists.

Consequently *the interior of a black hole is empty, with a singularity at the centre.*[11]

For those who already baulked at the extremely high mean densities of black holes the situation is still worse: all the mass of the black hole is theoretically enclosed in the central singularity, which has a mathematical volume of zero! Before we examine the problem of the central singularity, which has still not been resolved by modern physics, let us examine the region surrounding it.

This region moves because its geometry is collapsing. In other words, it is impossible to remain immobile in the interior of a black hole. This is illustrated in Figure 27. To remain immobile in this region would require a velocity in excess of that of light (the world-lines of the constant distance r are parallel to the time axis, and, inside a black hole, lie outside the light cones). Now the laws of relativity forbid motion faster than the velocity of light and this applies in the interior of a black hole as well as the outside. Inside the horizon the only allowed trajectories – confined inside the light cones – are ineluctably focused towards the central singularity.

A black hole can be compared to a 'world in reverse'. This image can be confusing but arises from the following analogy. In the region outside the black hole – for example the region of space-time which we inhabit – it is possible to move in a three-dimensional space in all directions, forward and back, left to right and up and down. However, time flows in only one direction, from the past to the future. It is a 'directed' coordinate whose flow is called *causality* (see page 23). Inside a black hole the roles are reversed. The coordinate which describes the distance to the centre of a black hole – which runs from $2M$ at the horizon to zero at the singularity – becomes the directing coordinate while the time coordinate becomes like the space coordinate outside the black hole. Inside the black hole, space becomes 'inexorable' in the sense that all matter is forced to watch its space coordinate diminishing, just as outside, all events have to move towards increasing times.

These concepts have to be treated with care. It does not mean that because the time coordinate inside the black hole becomes like

[11] The extrapolation is probably naive. It ignores the dynamical behaviour of matter inside a black hole; some speculations concerning this are developed in Chapter 19.

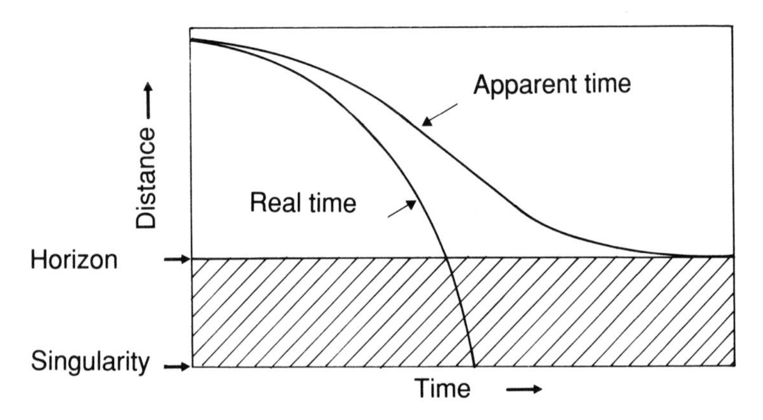

Figure 29: The two times of a black hole.

A finite interval of proper time elapses from the moment when the free-falling body crosses the horizon until it falls into the central singularity. However, to a distant observer, the approach to the black hole state appears to take an infinite apparent time; the body never crosses the horizon.

the space coordinate outside, it is possible to go back through time and violate causality. This coordinate changes its nature at the horizon and does not represent real time, either inside or outside the black hole (where it describes the apparent time measured by infinitely distant clocks). The only 'time' with a physical sense is the *proper time* measured by clocks free-falling towards the singularity. The proper time inside a black hole depends only on the distance coordinate to the centre, increasing as the latter decreases. Just as outside, true time always flows towards the future; here the only difference is that the future has an end – the singularity at the centre of the black hole. A finite interval of proper time elapses between the free-falling ship crossing the horizon and hitting the central singularity (Figure 29), whatever the power of the engines and the direction of navigation. This 'respite' time is longer for more massive black holes. For a 10 $M_\odot$ black hole it is just one thousandth of a second, but for giant black hole hidden in the heart of a galaxy the exploration could last an hour.

10

Illuminations

'The centre of the black hearth, of setting suns on the shore: ah! wells of magic.'

Arthur Rimbaud, *Illuminations*

A question of lighting

One of the best ways of representing a body is by its image, by photographing it. Can we imagine photographing a black hole?

The question is not as ridiculous as it seems, even though by definition a black hole does not emit light. This is true of all bodies which are sufficiently cold that they do not have their own detectable source of radiation. Everything has to be illuminated for us to able to see it. In the night sky, the planets, whose cores do not generate thermonuclear energy, would remain invisible if their surfaces did not reflect the Sun's light.[1]

In this respect what is true for a planet is also true for a black hole. A non-illuminated black hole would remain invisible, but with suitable lighting it is possible to obtain an image. A black hole can be photographed under lights!

Every natural body absorbs and reflects electromagnetic radiation in some way. The experiment illustrated in Figure 30 uses a parallel beam of light to illuminate several 'ideal' bodies, and to observe the reflected light in a direction perpendicular to the incident direction. The type of image received depends on the nature of the body and how it reacts to electromagnetic waves.

[1] Jupiter, the largest planet in the Solar System, possesses an internal source of energy. By contracting slightly, its core transforms atomic hydrogen into 'metallic' hydrogen, forming a solid lattice like ice. This 'phase transition' liberates a small amount of energy, providing Jupiter with its own luminosity, which is just greater than that coming from the reflection of solar light.

In the case of a perfectly *black* body (for example a sphere covered with perfectly absorbing black paint) all the light rays are absorbed, and in the absence of any reflection the observer will see absolutely nothing of the sphere (Figure 30*a*).

In the case of a *matt* surface (such as the Moon and the planets), the light is 'isotropically' reflected, that is, with the same intensity in all directions. Consequently at each point on the surface a single light ray may be deviated exactly 90° with respect to the incident direction and reach the observer. The result (Figure 30*b*) shows the familiar image of the half Moon.

The third example concerns a *perfectly reflecting* metallic sphere. This time a single point on the surface deviates an incident light ray by 90° and sends it to the observer. The image of the sphere (Figure 30*c*) is reduced to a unique spot of light, situated at a distance of 0.707 times the real radius of the sphere.

The final case is that of a *black hole*. The essential difference from the preceding examples is that a black hole is not a tangible hard surface against which light rays can strike and be reflected. It is the black hole's *gravitational field* which deviates the light rays. The black hole's sphere of influence is not therefore a single sphere – the event horizon – but extends to infinity. The trajectories of light rays are not altered by the impact with a surface, but are curved by the gravitational field. In the experiment where it is illuminated, the black hole's gravitational field deviates several light rays towards the observer. The image of the black hole consists of a series of illuminated spots (Figure 30*d*). On the left, at 2.96 times the Schwarzschild radius of the black hole, the 'primary' image is produced by light rays which have been deviated by 90°. On the right, 2.61 times the Schwarzschild radius the 'secondary' image is produced by light rays which have been deviated through another half circle (a total of 270°). The complete calculation of the geodesics of the Schwarzschild space-time corresponding to the trajectories of the light rays shows that there is an *infinity* of images; the third one corresponds to light which has been deviated by 450°, and so on adding a half turn each time. However, in practice, the images of order higher than two have such a low intensity and are so close to the primary and secondary images that they cannot be resolved.

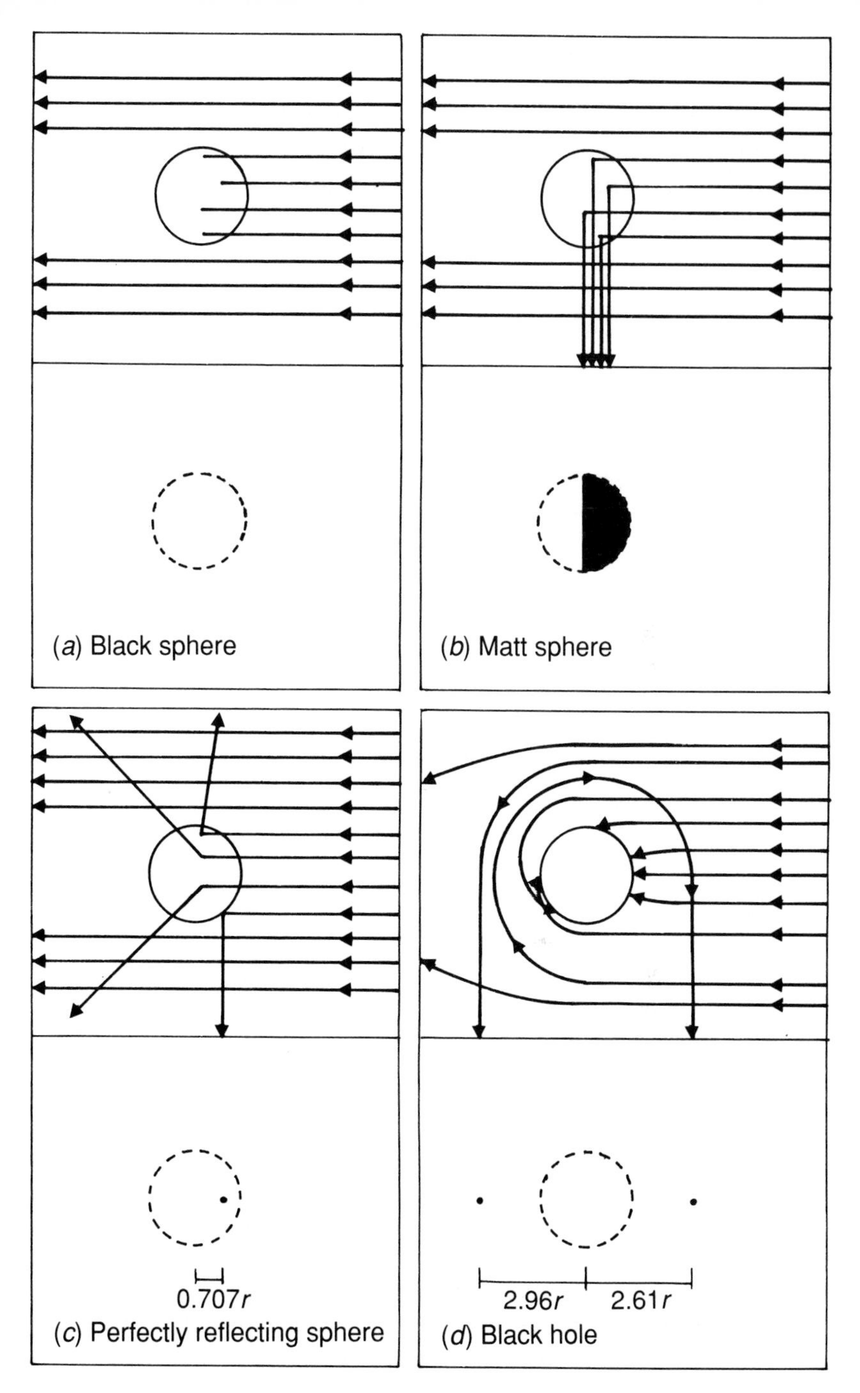

Figure 30: The illumination of four types of bodies.

We can therefore conclude that among the various celestial bodies which are not intrinsically bright, black holes are far from being the darkest. They are easier to detect than a black painted sphere or a highly reflective sphere!

The black hole in glory

A variant on the preceding experiment consists of illuminating a black hole and observing the light reflected back in the *same direction*. The result is shown in Figure 31.

The image of the actual black hole is magnified to 2.6 times its real diameter. This is because a large part of the incident beam is captured by the black hole: not just the radiation which directly strikes the event horizon but also that which passes within $5.2M$ of the centre (the true radius of the black hole is equal to $2M$). The black disc appears to be surrounded by a series of concentric light rings. The final result is like the *glory effect*, well known in traditional optics: when sunlight is scattered by innumerable water droplets in mist, it is sometimes possible to see in reflection the shadow of one's own head surrounded by brilliant rings of light centred on the line of sight.

In the case of black holes with a glory effect, only the outer ring would be visible, as it is not possible to resolve the rings of higher orders.

Heads and tails

The experiments which have just been described are more than an intellectual exercise, because if black holes really exist then there is a good chance that they will be illuminated by a natural light source.

For a black hole or a planet the most obvious form of lighting is a star. This star could, for example, be bound to the black hole in a binary system. Although such systems may be common throughout our Galaxy, the corresponding black holes would be impossible to detect by this effect as the image of the black hole by reflected

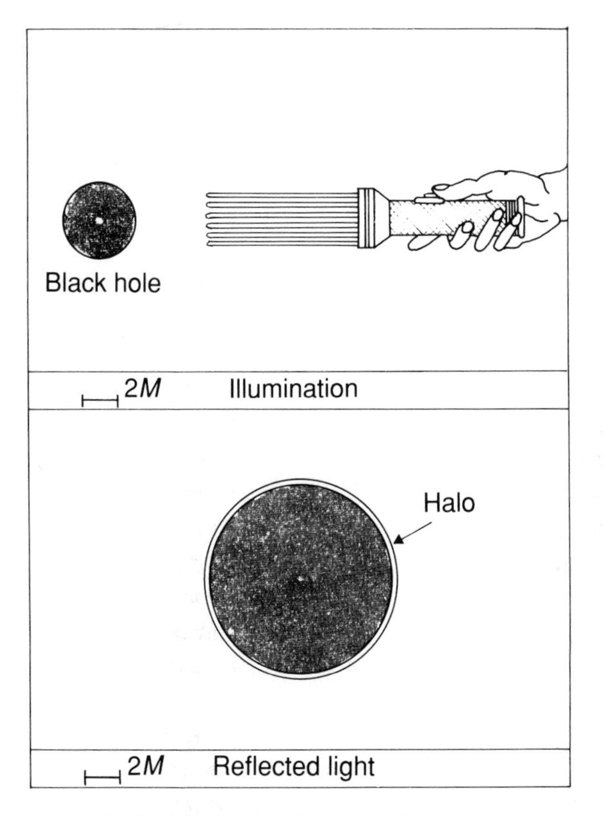

Figure 31. The halo of a naked black hole.

A black hole is illuminated by a parallel beam of light and one observes the returning light, formed of rays deviated 180° by the black hole's gravitational field.

light would be drowned in the intense light of the direct image of the star itself.

A much more interesting situation from an observational point of view is when the source of light comes from a series of rings of matter in orbit around the black hole. In Part 4, I shall give arguments for the belief that a number of black holes are surrounded by such structures, which are called *accretion discs*. Saturn's rings are an excellent example of an accretion disc; they consist of amalgamated pieces of rock and ice, whereas those of a black hole consist of hot gas.[2] The gases fall slowly into the black hole, like water in a whirlpool. As the gas falls towards the black hole it becomes hotter and hotter and begins to emit radiation. This is a source of light: the accretion rings shine and illuminate the central black hole.

Figure 32 shows the contours of a circular disc surrounding a spherical black hole. The image was photographed at a great distance, in a direction which was slightly inclined above the plane of the disc. The strong curvature of space-time near the hole manifests itself as a distortion of the apparent image of the disc. In the case of Saturn's rings, this is reduced to a series of ellipses because they are seen through a practically flat space-time. Here, the image is split into two. The *primary image* is formed by light rays emitted from the *upper side* of the disc, and deviated by less than 180°. The first surprise is that *all of the upper side of the ring in front and behind the black hole is visible*, including the portion which would 'normally' be hidden in an uncurved geometry (Saturn's rings observed from the Earth are partially hidden by the planet's disc).

Even more surprising is that the curvature of space-time around the black hole enables us to observe the *lower side* of the rings. This is the *secondary* image. So it is possible to observe both the upper and lower sides of the accretion disc!

In fact there is an infinity of images, because the light rays emitted by the disc can travel any number of times around the black hole before escaping from its gravitational field and being observed by a distant astronomer. The primary image shows the upper side and the secondary image shows the lower side, the third image shows the upper side again, and so on. However, higher order images are not optically interesting because they are stuck to the edge of the central black disc, the latter representing an expanded image of the actual black hole.

2 Another important difference is that the accretion disc of a black hole is continually being supplied with gas, whereas that surrounding Saturn is the remnant of the primordial Solar System.

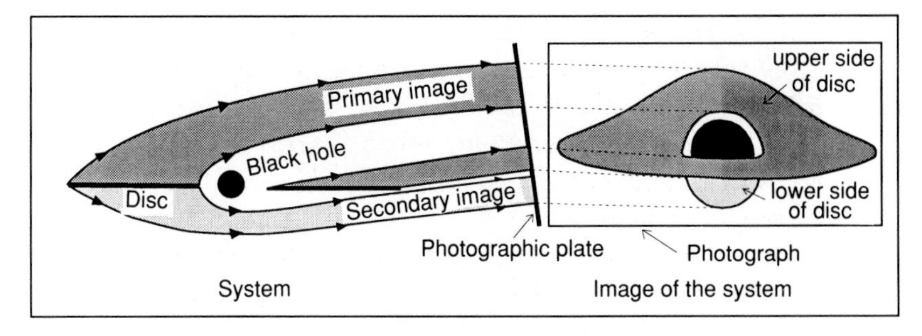

Figure 32. Optical distortions near a black hole.

We imagine a black hole surrounded by a bright disc. The system is observed from a great distance at an angle of 10° with respect to the plane of the disc. The light rays are received by a photographic plate. Because of the curvature of space-time in the neighbourhood of the black hole, the image of the system is very different from the ellipses which would be observed if an ordinary star replaced the black hole. The light emitted from the upper side of the disc forms a direct image and is considerably distorted so that it is completely visible. The lower side of the disc is also visible as an indirect image caused by highly curved light rays.

Photographing a black hole

These illumination experiments, although idealised, at least show how a black hole, through its gravitational field, acts on radiation as a sort of lens which *multiplies* images of a single unique source. Let us now consider a more realistic situation. Rings of matter around celestial bodies have been intensively studied over the last 20 years because they are related to a large number of astronomical phenomena: the planets (Saturn, Jupiter, Uranus and Neptune), and binary star systems where one of the partners is a condensed star (white dwarf, neutron star and black hole). The gravitational field around a black hole sucks in the overflowing gas of its companion, stores it in the accretion disc and slowly consumes it.

Detailed models of accretion discs explain the high energy radiation emitted by some binary star systems such as Cygnus X-1. On a much greater scale, the high luminosity of the nuclei of

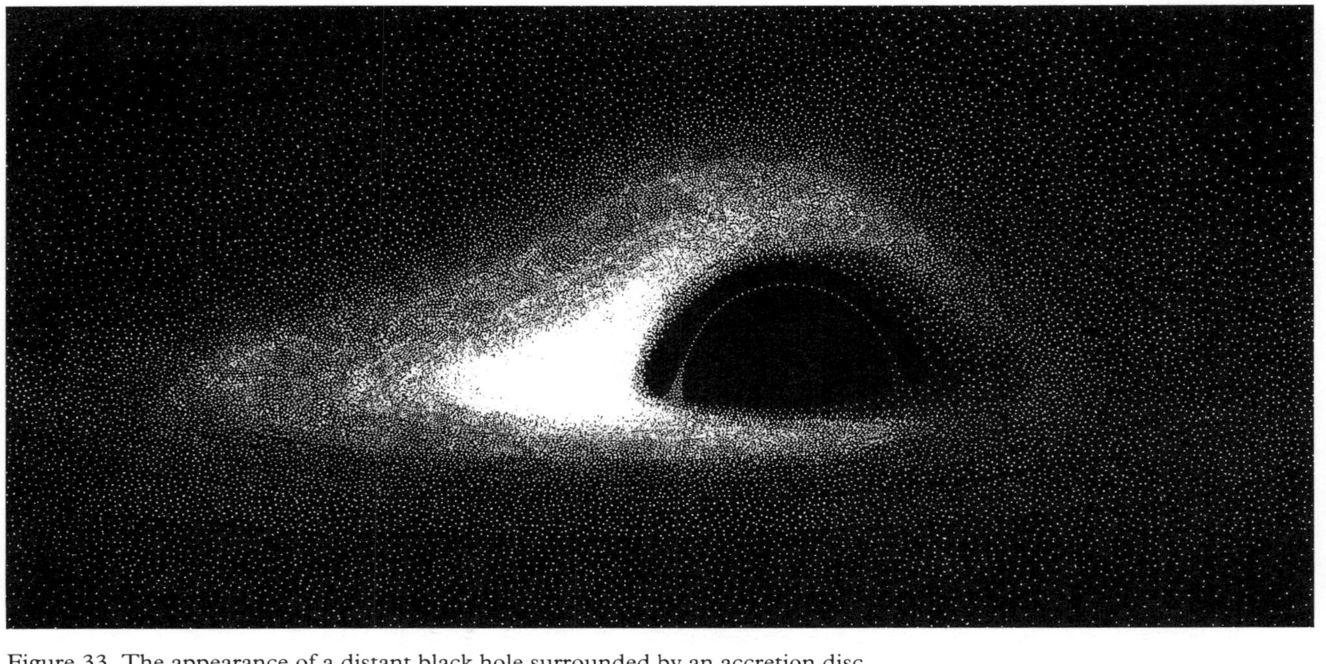

Figure 33. The appearance of a distant black hole surrounded by an accretion disc.

The image was calculated by a computer. As in the preceding image, the system is seen from a great distance, inclined at an angle of 10° above the plane of the disc. The image is realistic in the sense that it takes account of the physical properties of the gaseous disc.

certain galaxies and quasars can also be explained by the flow of matter into a black hole of several million or several billion $M_\odot$. A detailed discussion of the role of black holes in these various astronomical phenomena will be the subject of Part 4. For the moment it is enough to know that when the flow of matter into a black hole is not too great, it can form a *very thin* accretion disc, whose emission of radiation can be accurately calculated.

In 1978, I reconstructed the photographic appearance of a spherical black hole surrounded by a very thin gaseous disc, by using a computer to calculate the trajectories of light rays in the Schwarzschild space-time (Figure 33). In a thin disc the intensity of radiation emitted from a given point on the disc depends only on its distance from the black hole. The reconstituted image is therefore universal, that is, it is independent of the mass of the black hole and the flow of gas into it. It could as easily represent a black hole 10 kilometres in diameter or one that was as large as the Solar System and swallowing interstellar gas.

As in Figure 32, the upper side of the disc is completely visible. However, only a small part of the lower side is observable. In a realistic situation the gaseous disc is opaque; therefore it absorbs the light rays that it intercepts. It follows that a major part of the secondary image – showing the lower side of the disc – is occulted by the primary image; the very deformed visible part is at the edge of the black hole.

No radiation comes from the region between the black hole and the inner edge of the disc. The properties of the Schwarzschild space-time forbid the accretion disc from touching the surface of the black hole. The almost circular orbits of the gas in the disc can be maintained only down to a critical distance of three times the Schwarzschild radius. Below this the disc is unstable; the gas particles plunge directly towards the black hole without having enough time to emit electromagnetic radiation.

The main characteristic of a 'photograph' of a black hole is the apparent difference in luminosity between the different regions of the disc. The maximum brilliance comes from the inner regions closest to the horizon, because it is here that the gas is hottest. However, the apparent luminosity of the disc is very different from the intrinsic luminosity. Apart from the geometric distortion of the circular rings, the radiation picked up at a great distance by a photographic plate is frequency- and intensity-shifted with respect to the emitted radiation. There are two sorts of shift effects. There is the *Einstein effect*, which has been mentioned several times, in which the gravitational field lowers the frequency and decreases the intensity. In addition there is the better known *Doppler effect*, where the displacement of the source with respect to the observer causes amplification as the source approaches and attenuation as the source retreats.[3] In this case the Doppler effect is caused by the disc rotating around the black hole. The regions of the disc closest to the black hole rotate at a velocity approaching that of light, so that the Doppler shift is considerable. The sense of the rotation of the disc is such that matter recedes from the observer on the right-hand side of the photograph and approaches on the left-hand side. As the matter recedes, the Doppler deceleration is added to the gravitational deceleration, explaining the very strong attenuation on the right-hand side. In contrast, on the left-hand side the two effects tend to cancel each other out, so the image more or less retains its intrinsic intensity.

[3] Also see Chapter 16.

11

A descent into the maelstrom

'I became possessed with the keenest curiosity about the whirl itself. I positively felt a wish to explore its depths, even at the sacrifice I was going to make; and my principal grief was that I should never be able to tell my old companions on shore about the mysteries I should see.'

Edgar Allan Poe, 'A descent into the maelstrom' (1840)

The Kerr black hole

All the stars rotate. For this reason they are not exactly spherical, but slightly flattened at the poles. The gravitational collapse of a real star is not therefore exactly described by the spherically symmetric Schwarzschild solution. In reality, the geometry of the surrounding space-time will become much more complicated because of the production of *gravitational waves*.

Why do these gravitational waves[1] disturb the geometry? The reason is simple: all moving matter (for example a rotating star) has a gravitational field which varies with time. Consequently, the curvature it causes in space-time changes at each instant to reflect the new configuration of matter. These readjustments propagate at the velocity of light as 'wrinkles' in the curvature, travelling across the background geometry.

Collapsing stars with the least spherical shapes emit the most gravitational waves. If the star collapses into a black hole and an event horizon is formed then the situation simplifies itself in an instant. At the moment of its formation the horizon may still have

[1] See Chapter 18.

an irregular shape and be subjected to violent vibrations; however, a fraction of a second later the gravitational waves smooth out all irregularities (Figure 34). The horizon stops vibrating and assumes a unique smooth shape: a spheroid flattened at the poles by centrifugal forces.

This is why the gravitational field of a rotating star that collapses into a black hole reaches a final state of equilibrium which depends on two parameters only: mass and *angular momentum*. The latter is related to the rotational motion of the star and is similar to the spin of elementary particles (see page 73).

There is an exact solution of Einstein's equations which depends only on these two parameters. It was discovered in 1962 by the New Zealand physicist Roy Kerr and describes the gravitational field of a rotating black hole. The astronomical implications of this theoretical discovery were considerable, comparable to the discovery of a new elementary particle. Science has always been like this, theory and experiment feeding each other.

Whereas the Schwarzschild geometry describes a gravitational field caused by a spherical mass, static or not, Kerr's geometry describes a final equilibrium solution which can only be applied once the event horizon has formed and all the distortions have been 'swept away' by gravitational waves, and not during the actual collapse of the rotating star.

The maximal black hole

Most stars are in differential rotation. They are composed of layers of gas of varying densities, which do not rotate at the same velocity. In the Solar System, the atmospheres of the gaseous planets such as Jupiter and Saturn show the effects of differential rotational by exhibiting elongated bands parallel to the equator. The Kerr black hole is rotating with perfect *rigidity*: all the points on the horizon move with the same angular velocity.

On the other hand, stars cannot rotate with just any velocity. Even neutron stars, which are like giant spinning tops, cannot perform more than a thousand rotations per second: above this limit they would disintegrate under centrifugal forces. There is a critical angular momentum beyond which the event horizon would

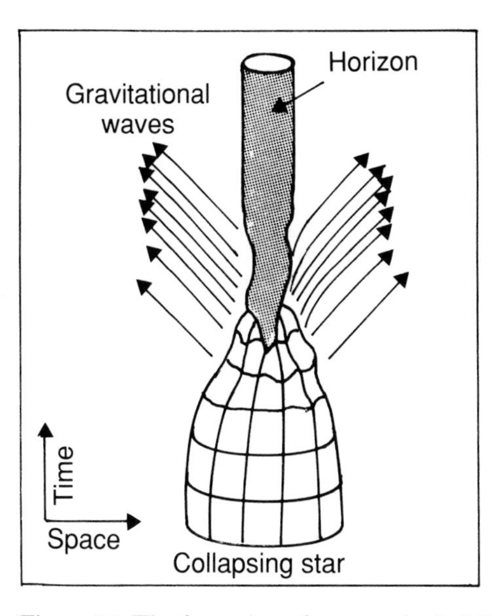

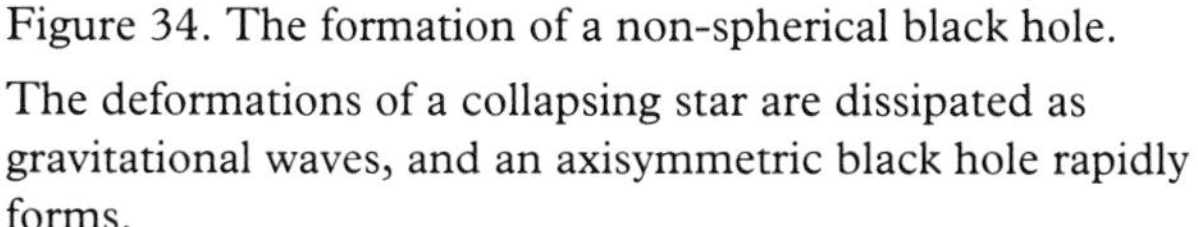
Figure 34. The formation of a non-spherical black hole. The deformations of a collapsing star are dissipated as gravitational waves, and an axisymmetric black hole rapidly forms.

'break up' leaving the naked central singularity. This limit corresponds to the horizon having a rotational velocity equal to that of light. For such a black hole, called a 'maximal' black hole, the gravitational field at the event horizon would be zero. In Newtonian language this means that on the event horizon the force of centrifugal repulsion exactly cancels the force of gravitational attraction.

It is possible that most of the black holes formed by the gravitational collapse of a massive star have an angular momentum which is very close to this critical limit. In fact, a number of rotating stars, although far from being black holes, already have a very high angular momentum (the Sun's is 20% of the critical limit). If the angular momentum is conserved during the collapse,[2]

[2] The conservation of the angular momentum explains the very high rotational velocities of neutron stars, see Chapter 7.

it is likely that stellar black holes approach this limiting state. Thus black holes of 3 $M_\odot$, believed to be the 'engines' of binary X-ray sources (see Part 4), must rotate at almost 5000 revolutions per second.

But a black hole is not a spinning top revolving in a fixed exterior space. We cannot fix a lamp on the horizon and count the number of passes per second. As it rotates, a Kerr black hole *drags the entire fabric of space-time along with it.*[3] Theoretically it is only at an infinite distance that space-time ceases to 'rotate' and that it is possible to attribute an angular velocity to the horizon. Closer to the black hole, space-time is irresistibly sucked into a whirlpool shape. After the capture of light, this is the second fundamental characteristic of a black hole: *it is a cosmic maelstrom.*

The cosmic maelstrom

'But little time will be left to me to ponder upon my destiny! The circles rapidly grow small – we are plunging madly within the grasp of the whirlpool – and amid a roaring and bellowing and thundering of ocean and of tempest, the ship is quivering – oh! God! – and . . . is going down!'

Edgar Allan Poe, 'Manuscript found in a bottle'

There is a profound analogy between a rotating black hole and the familiar phenomenon of a vortex, whether it be the swirling water which goes down a plug hole when a bath is emptied, or the giant whirlpools produced by sea currents, such as the legendary maelstrom off the coast of Norway (described by Edgar Allen Poe in his *Tales of the Grotesque and Arabesque*), or even the Corrievreckan in the Scottish Hebrides mentioned by Jules Verne in his book *The Green Ray.*[4]

In a whirlpool, water moves in a spiral which can be decomposed

[3] General Relativity states that it is also true for all massive rotating bodies, but the dragging of the geometry, called the *Lense–Thirring effect*, is minimal unless the body has collapsed into a black hole.

[4] Also it should not be forgotten that at the end of *Twenty Thousand Leagues under the Sea*, Jules Verne caused the submarine *Nautilus* to disappear into one of these marine abysses.

into a circular motion and a suction towards the centre. The circular motion has a purely tangential velocity proportional to the inverse square of the distance from the centre of the whirlpool; the suction has a purely radial velocity which is much smaller than the tangential velocity and varies as the inverse of the distance from the centre.

Now imagine that a motor boat ventures into a whirlpool (Figure 35). We assume that the boat has a maximum velocity of 20 km/h in calm water. Far from the whirlpool the captain can obviously navigate where he wants; the motor has no problem in overcoming the motion of the water. He can thus remain in a fixed position without needing to anchor. He can approach or move away from the whirlpool, and travel against the current.

If the captain decides to navigate towards the whirlpool, there will come a point at a certain distance from the centre where the circular velocity of the current equals the maximum velocity of the boat, 20 km/h. Inside this critical distance, the boat is unable to maintain a fixed position, even with its engine running at full speed. It is forced to travel in the same direction as the rotation of the whirlpool. In other words, the boat which was originally free to travel in any direction is now limited to a region bounded by the inside of an angle closed by the straight lines joining the boat position and the tangents to the 'navigation circle' in front of the boat. However, although dragged along by the circular current, the boat can still escape from the whirlpool by orienting itself on a suitable trajectory and spiralling outwards.

If the boat travels too close to the centre of the whirlpool there will come a fatal moment when the radial velocity of the current equals 20 km/h (the circular velocity is already much greater). The navigation circles then plunge directly into the mouth of the whirlpool; as Edgar Allan Poe wrote 'if a ship comes within its attraction, it is inevitably absorbed and carried down to the bottom, and there, beat to pieces'.

The analogy with the Kerr geometry surrounding a rotating black hole is clear. The centre of the whirlpool is the black hole. The water surface pulled down by the whirlpool is the space-time curved by gravitation and drifting in the sense of the vortex. The boat could be a spaceship or any material particle, whose

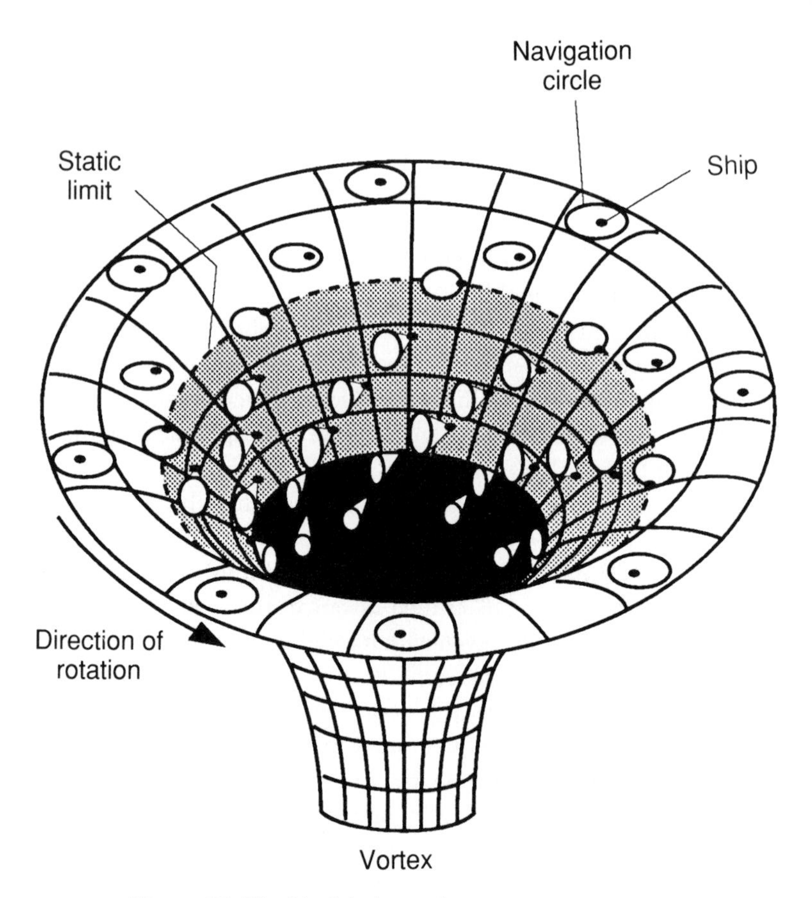

Figure 35. The black hole maelstrom.

The gravitational well caused by a rotating black hole resembles a whirlpool. A spaceship travelling in the vicinity of a black hole is sucked towards the centre of the vortex like a boat. In the region outside the static limit (clear), it can navigate where it wants. In the zone (grey) between the static limit and the event horizon it is forced to rotate in the same sense as the black hole; its ability to navigate freely is decreased as it is sucked inwards, but it can still escape by travelling in an outwards spiral. The dark zone represents the region inside the event horizon; any ship which ventured there would be unable to escape even if it was travelling at the velocity of light.

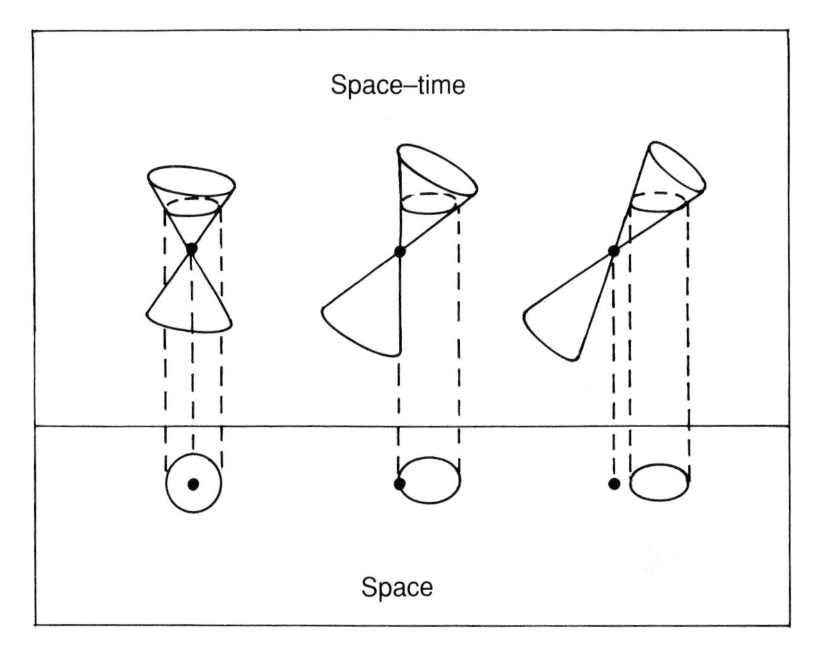

Figure 36. Navigation circles.

If we cut a light cone at fixed time (a horizontal plane in this figure) the resulting spatial section is a 'navigation circle' (or more exactly an ellipse) which determines the limits of the permitted trajectories. If the cone tips over sufficiently in the gravitational field, the navigation circle detaches itself from the point of emission. The navigational directions are confined within the angle formed by the tangents of the circle and it is impossible to return.

maximum allowed velocity is that of light, 300 000 km/s. As Figure 36 shows, the navigation circle at a given point is a spatial projection of a light cone which marks out the allowed trajectories.

The light cones are not only deviated towards the interior of the gravitational field, but are also dragged in the rotational sense of the black hole. This 'spiral' is inexorable inside what is called the *static limit.* In this region, the circles of light – the projections of the cones – are detached from their emission point and shifted forwards. Consequently the spaceship is unable to remain static with respect to a distant fixed reference frame (for example the stars), even if it is travelling at the velocity of light.

Closer still to the centre of the black hole is a second critical surface where the light cones are tipped so far inwards that nothing can escape from it. We recognise the *event horizon*, the true boundary of the Kerr black hole.

The event horizon is situated inside the static limit, but these two characteristic surfaces of the Kerr black hole touch at the poles (Figure 37). They have quite distinct roles. Time appears to 'freeze' at the static limit and radiation acquires an infinite redshift, but it is only at the event horizon that matter is completely imprisoned.[5]

The region of space-time between the two surfaces is called the ergosphere. The name was invented by John Wheeler from the Greek word for 'work', because it is theoretically possible to use some of its unusual properties to extract the rotational energy of the black hole. I will return to this astonishing speculation in Chapter 13.

The ring singularity

The internal structure of a rotating black hole is much more complex than that of a static black hole. The first important difference is the central singularity, where the curvature becomes infinite. In a rotating black hole it is no longer a point but a flat *ring* in the equatorial plane. This ring is no longer a space-time knot towards which all matter must converge. It is now possible to travel inside a rotating black hole by avoiding the ring singularity, either by travelling above its plane or by passing through it! These new possibilities for exploring black holes will be discussed in Chapter 12.

There is another difference: there is a second event horizon inside the actual boundary of the black hole. This spherical surface surrounds the ring and 'protects' the region between the internal and external event horizon from the singularity.[6] As the angular

[5] In the Schwarzschild black hole we recall that the single horizon has both properties.

[6] In the sense that a signal emitted by the singularity cannot escape from the internal horizon.

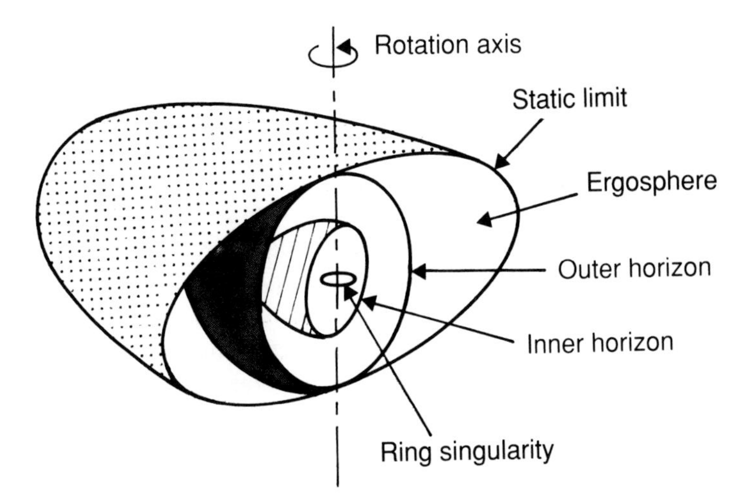

Figure 37. Cross-section of a rotating black hole. This spatial representation illustrates the complex internal structure: multiple horizons and a ring singularity.

momentum of the black hole increases, the two horizons tend to merge, the internal horizon expands and the external one contracts. At the limit, for a maximal black hole rotating at the critical speed, the two horizons break up leaving a naked gravitational singularity.

The electric black hole

Stars which collapse into black holes generally possess a magnetic field. In addition, black holes swallow electrically charged particles from the interstellar medium, such as electrons and protons. It is therefore reasonable to expect black holes to have electromagnetic properties.

H. Reissner in 1916, and independently G. Nordstrøm in 1918, discovered an exact solution to Einstein's equations for the gravitational field caused by an electrically charged mass. This solution is a generalised version of Schwarzschild's solution, with one other parameter: the electric charge. It describes space-time outside the event horizon of an *electrically charged black hole.*

If the electromagnetic properties of a black hole reduce to a single electric charge, the electromagnetic structure of the parent star (field lines, existence of magnetic poles and so on) must have simplified considerably. Here again, gravitational waves have carried away most of the electromagnetic attributes of the star, leaving behind only a global electric charge, not localised on the horizon, analogous to the electric charge of an elementary particle. This charge does not alter the shape of the black hole, which remains perfectly spherical in the absence of rotation.

There is a limit to the amount of electric charge a black hole may have. Above a critical limit the event horizon would be destroyed by the colossal force of electrostatic repulsion. The maximum electric force is proportional to the mass of the black hole, and for a 10 $M_\odot$ black hole it is 10^{40} times the charge on an electron. Nevertheless, a black hole is just as likely to be positively charged as negatively charged.

The interior structure of a highly charged black hole has features common to that of a static neutral hole or a rotating hole; as in the former, the singularity is a point, but as in the latter, it is screened by an inner event horizon. The area of the inner horizon increases and that of the outer horizon decreases as the electric charge grows. At the maximum permitted charge, the two horizons would merge together and disappear, revealing the singularity to distant astronomers.

Despite these subtleties, this discussion of highly charged black holes is fairly academic, because 'natural' black holes are probably neutral. This is for the same reason that most ordinary matter is neutral: the remarkable weakness of gravitation compared with electromagnetic interactions. A macroscopic body (that is, one containing an enormous number of elementary particles) contains almost exactly equal numbers of positive and negative charges (carried by electrons and protons). The electrostatic forces cause these charges to associate and neutralise each other. Let us now imagine that a black hole has formed with a large positive charge, near to the maximum value allowed. In a realistic astrophysical environment, the black hole will not sit in a complete vacuum but in the interstellar medium, which is full of protons and electrons. The black hole's gravitational field will attract both electrons and protons; however, its electric charge will attract only charges of the opposite sign, electrons, and will repel the protons. The electrostatic forces are a billion billion times stronger than the gravitational forces. Therefore in a very short time the black hole will have captured all the available electrons and will have almost completely neutralised itself. The electric charge of a 'natural' black hole cannot be any greater than one billion billionth of its maximum charge. This is so small that the astrophysical effects of the black hole's electric charge can be ignored.

A black hole has no hair

Are there as many types of black holes in the Universe as there are stars? In other words, apart from mass, angular momentum and electric charge, what parameters can black holes have?

For a physicist a star or a sugar cube are fantastically complicated objects, in the sense that a *complete* description of them, including their atomic and nuclear structure, requires billions of parameters. But a physicist studying the exterior of a black hole does not have the same problems. A black hole is by contrast an incredibly simple object: if we know its mass, angular momentum and electric charge, then we know *everything* there is to know about it.

A black hole retains practically nothing of the complexity of the matter which formed it. It does not remember its shape or composition; it keeps only the mass, angular momentum and electric charge (Figure 38). This disarming simplicity is perhaps the most basic characteristic of a black hole. John Wheeler, who invented most of the terminology concerning black holes, remarked in the 1960s that 'black holes have no hair'.

What began as conjecture has recently received a mathematically rigorous proof, a result of efforts over 15 years by half a dozen theoreticians, including Brandon Carter of the Observatoire de Meudon and the Australian Gary Bunting. Confirming Wheeler's statement, their work showed that only three parameters were needed to describe the geometry of space-time around a black hole *in equilibrium*. For the theoretician this implies a considerable simplification: there are only four types of black hole, depending on which parameter is the most important.[7] Let us list them: the spherical and static Schwarzschild black hole characterised by its mass; the Reissner and Nordstrøm black hole, which is also spherical and static but has an electric charge; the Kerr black hole, which is a rotating neutral mass; and finally the most general equilibrium black hole, rotating and charged, which was calculated in 1965 and given the name Kerr–Newman. This last solution represents the *unique and natural final state of a gravitational collapse within an horizon.* As we have explained, the electric charge plays a negligible role, so the most 'realistic' black hole is correctly described by the Kerr solution.

Once again gravitational waves sweep away all the complex structure of the matter as the black hole forms. They 'shave the hair' from the black hole leaving only its mass, angular momentum and electric charge. These physical parameters are characteristic of the two long-range interactions present at the formation of a black hole: gravitation (for the mass and angular momentum) and electromagnetism (for the electric charge). The short range nuclear interactions which structure atomic nuclei play no role in the formation of black holes.

The black hole parameters are perfectly measurable, albeit by means of thought experiments. A black hole could be weighed by placing a satellite in orbit around it and measuring its orbital period. Its angular momentum could similarly be measured by comparing the deviations of light rays sent towards various parts of its horizon.

For a general Kerr–Newman black hole of given mass, the electric charge and the angular momentum both have upper limits. They are constrained by a relation ensuring the existence of the event horizon. If this constraint is violated – for example by the gravitational collapse of a massive star – the black hole would become a naked singularity, capable of influencing the Universe at large distances. Physicists have good reason to believe that such a situation is prevented by the laws of Nature.[8]

[7] The mass which is responsible for the gravitational field is of course taken into account.

[8] This important subject will be dealt with in the next chapter.

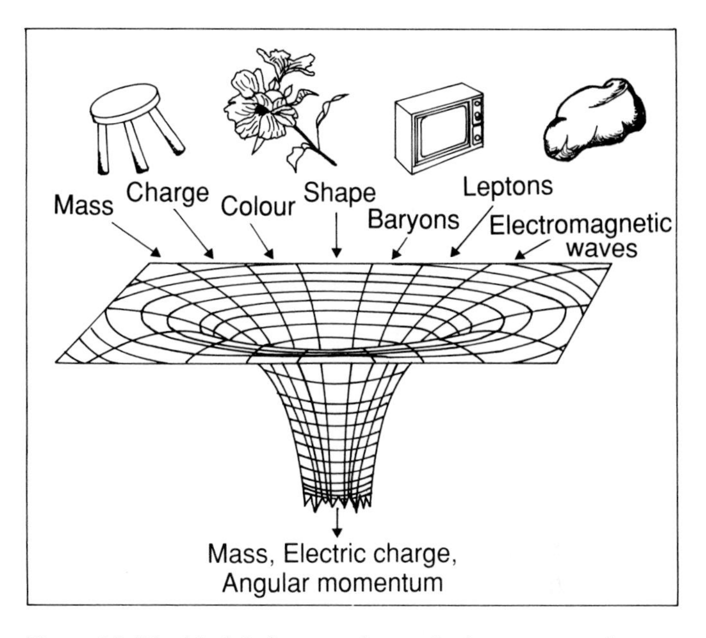

Figure 38. The black hole remembers only the mass, angular momentum and electric charge of the matter which falls into it. (After Ruffini and Wheeler.)

Since it is only affected by three parameters, a black hole is of the same order of simplicity as an elementary particle. However, if we examine the condition of the existence of an event horizon, nothing could be more unlike a black hole than an elementary particle, even though the latter assembles mass, angular momentum and charge in a very small volume. Taking the case of an electron, we know from experiment its mass, angular momentum (spin) and electric charge. For its given mass, the electric charge and angular momentum of an electron exceed the black hole limits by a factor of 10^{88}. This staggering number, which is even greater than the total number of elementary particles in the observable Universe, is a measure of the difference between an electron and a Kerr–Newman black hole![9]

[9] Which is not to say that an electron is a naked singularity!

12

Map games

'The map is not the territory.'

Alfred Korzybski

Black and white

The human mind has a natural preference for symmetry. Since antiquity, physicists have been trying to analyse the mechanisms of Nature in terms of elementary symmetry. The surprising thing is that this method has often been successful. An excellent example of this was the theoretical prediction of 'anti-particles', followed shortly after by their experimental discovery. Symmetry is more important now than ever in the most recent developments in fundamental physics.

The black hole has a symmetrical opposite, the *white hole*, a sort of gravitational outflow from a region hidden behind a horizon. Early interpretations of the white hole led to the popular idea that man could travel instantaneously from one part of the Universe to another by entering a black hole and reappearing from a white hole, having travelled through the 'throats' connecting them. Such ideas certainly increased the fascination of black holes for the general public, but reduced their credibility amongst scientists unfamiliar with General Relativity.

What is the real status of white holes? We have to re-examine the delicate problem of the relationship between the real world and its mathematical description, or between a map and the actual territory. One of the most common symmetries in the laws of physics is time reversal. In Galileo and Newton's mechanics, Fresnel's optics, Maxwell's electromagnetism and Einstein's relativity, all the equations are symmetric with respect to time.

Thus from a given set of conditions at a certain instant in time it is possible to calculate the trajectories of a planet, a light ray or an electron into both the future and the past. This does not mean that Nature is indifferent to the flow of time; the light rays actually leaving the surface of the stars travel into the future and not the past.

In other words, the solutions to the 'equations of physics' do not necessarily occur in the real world. It is not, however, always simple to sort out the real solutions from the fictional ones. In particular, great care is needed when considering the physical interpretation of symmetrical solutions, even if they are aesthetically attractive.

Dennis Sutton has written that, 'the frontiers of science are always a bizarre mixture of new truth, reasonable hypotheses and wild conjecture'. This quotation applies well to this book: at present we can say that General Relativity belongs to the first category, black holes to the second and white holes to the third. However, it is also fair to say that some of the 'wildest' conjectures have helped to advance science, and for this reason white holes are worth some attention. Their charm is increased by the fact that their study has a playful aspect to it, not usually expected by the general public, but essential for many scientists. So, let's play!

The 'embedding' game

When trying to understand abstract notions, a problem that comes up time and time again is that of visualisation. Taking the example of space-time, the analogy between a 'space-time mollusc' locally curved by matter and the 'rubber band' of space deformed by marbles enables us mentally to represent certain aspects of the curvature of abstract four-dimensional geometries. This description can be made rigorous by a mathematical technique called *embedding*.

As the name suggests, the game is to visualise the shape of a given space by embedding it in a space with an additional dimension. For example, the shape of a circle (one dimension) is easy to visualise by embedding it into a plane (two dimensions), and the surface of a sphere (two dimensions) is easy to visualise in ordinary Euclidean space (three dimensions).

The embedding game is useless for the full four-dimensional space-time continuum, as it has to be embedded in a five-dimensional space, which is impossible to imagine.[1] Fortunately there are a number of variations of this technique which can be used.

Let us assume, for example, that space-time is *static*, that is, that the spatial geometry remains fixed at all times. In this case no information is lost if we visualise it as *instantaneous time-slices*. Furthermore, if the spatial geometry is spherical, no information is lost by looking only at *equatorial slices* passing through the centre of the sphere. Therefore it is easy to cut spherical static space-time into two-dimensional slices without losing any information about the curvature of the complete space-time manifold. It then becomes easy to visualise all the details of the curvature by embedding it in a three-dimensional Euclidean space.[2]

Let us now consider the practical implications of this in the case of space-time deformed by a spherical star in equilibrium, such as the Sun. Since the geometry is static both outside and inside the star, the instantaneous equatorial slices all have the same form, represented by the curved surface in Figure 39.

The shape of this surface is reminiscent of an elastic sheet deformed by the weight of a marble. It is divided into two regions. The zone extending to infinity represents space-time outside the star. It is a region of Schwarzschild geometry. As to the zone occupied by the star, its exact shape depends on the internal structure of the star but remains similar to that of a portion of a sphere. Since the star is not collapsed, the Schwarzschild critical radius $r = 2M$ lies inside the star and there is no central singularity: the hollow has a perfectly regular curvature.

This type of representation is both informative and rigorous, and has already been used in Figure 16 to represent the curvature of light rays passing near the Sun.

[1] It is even mathematically impossible to embed a four-dimensional space-time into five-dimensional Euclidean space.

[2] This is just a fictitious space which serves to 'contain' the space-time sections.

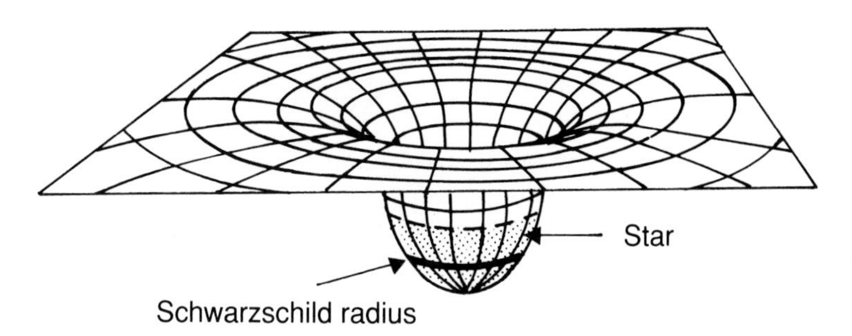

Figure 39. Embedding of a non-collapsed spherical star.

The embedded surface shows the curvature of space around the star. All points situated outside the embedded surface are without physical significance. Each circle traced on the surface represents the set of points situated at the same distance from the centre of the star, while the orthogonal curves pass through the bottom of the hollow which is the centre of the star. At large distances from the star the gravitational field is weak and the embedding surface loses its curvature. However, it does not become a horizontal plane as the figure suggests, but rather a paraboloid. Near the star the curvature is more accentuated. The shaded area indicates the region effectively occupied by the star.

Wormholes

'These sedentary forms build temporary or permanent tunnels. The sandworm lives in a simple U-shaped gallery.'

Encyclopaedia Universalis, 'Annelids'

Let us now apply the embedding technique to the case of a spherical black hole. Figure 40 shows that there is a surprise in store: the embedding surface consists of a paraboloid-shaped[3] *throat* linking two distinct symmetric sheets of space-time. How do we interpret this unexpected shape? Unlike the case of an ordinary star, only the space-time manifold outside the black hole can be represented. The throat has a minimum radius

[3] The surface generated by rotating a parabola about its axis of symmetry.

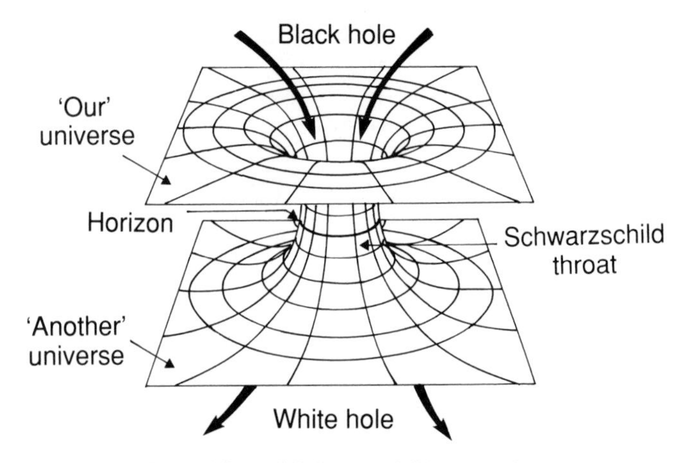

Figure 40. Embedding of Schwarzschild space-time.
The Schwarzschild throat connects 'our' universe (upper sheet) with 'another' universe (lower sheet).

equal to the Schwarzschild radius $r = 2M$. It is therefore the event horizon, the frontier of the black hole reduced to a circle.

Forget for a moment the double structure of the embedding surface and concentrate on the upper sheet (Figure 41). It extends to infinity and slowly loses its curvature as the distance from the throat increases: it is asymptotically flat. The trajectories of free-falling particles and light rays are 'straight' lines traced on the curved surface, that is, *geodesics*. These become more curved as they pass closer and closer to the gravitational well. Some of them plunge so deep into the well that they intersect themselves at the exit; those which encounter the central circle of the throat, the horizon, can never escape.

Figure 42 projects the preceding geodesics onto the plane (P) parallel to the horizon circle. The result illustrates the Principle of Equivalence perfectly, giving the Newtonian illusion of a flat universe in which particles are deviated from a straight line by gravitational 'forces' instead of freely following the contours of the underlying curved geometry.

We now return to the embedding surface in its entirety (Figure

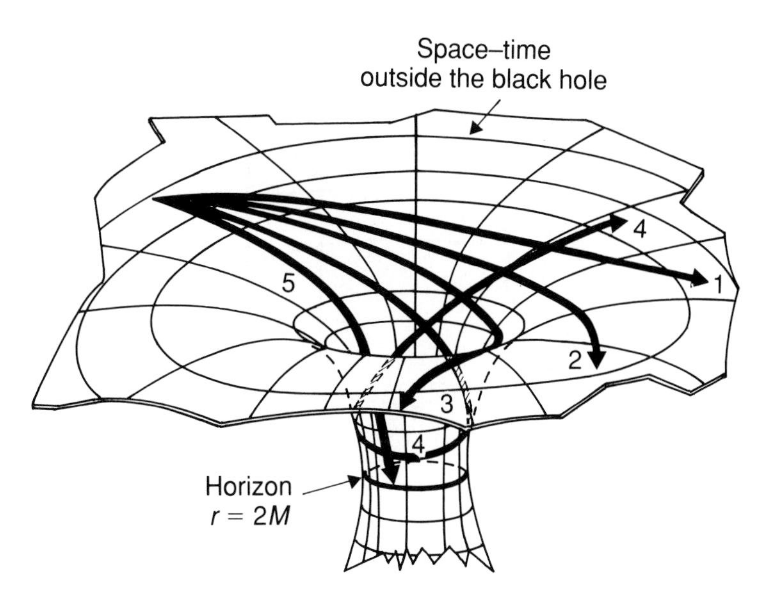

Figure 41. Geodesics in Schwarzschild space-time.

The five geodesics drawn on the embedding surface represent the possible trajectories of a free-falling body passing at different distances from a black hole. The geodesics 1, 2 and 3 are increasingly affected by the curvature. Geodesic 4 falls into the well and intersects itself as it comes out. Geodesic 5 falls straight into the hole and does not reappear.

40). The Schwarzschild throat[4] joins the upper and lower sheets which are perfectly symmetrical and asymptotically flat, and which we are at liberty to interpret as 'parallel universes'. A geodesic entering the throat from the upper sheet appears to be able to leave via the lower sheet. In other words, the Schwarzschild throat appears to the upper universe like a black hole consuming matter, but for the lower universe it appears as an 'anti-black hole' expelling matter. It does not require much imagination to name this anti-black hole a *white hole*, or more correctly a *white fountain* (reversing both the noun and the adjective).

The embedding game becomes even more disconcerting if we remember that General Relativity determines only the *local*

[4] Also called the 'Einstein–Rosen bridge'.

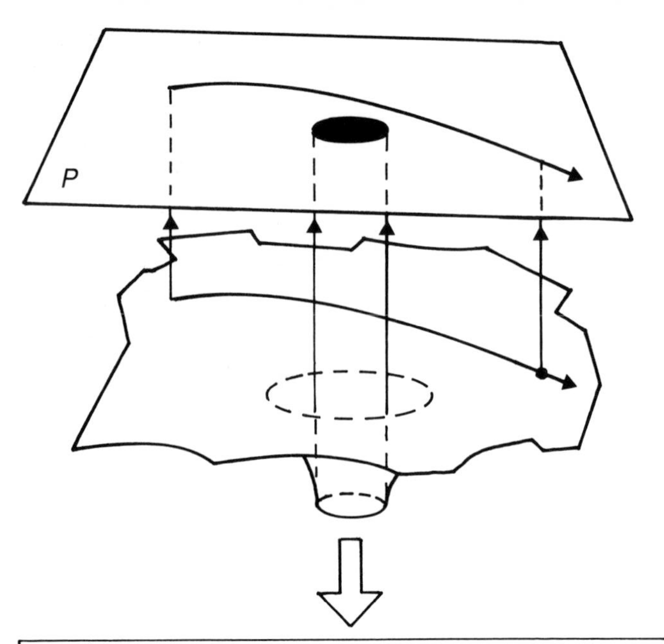

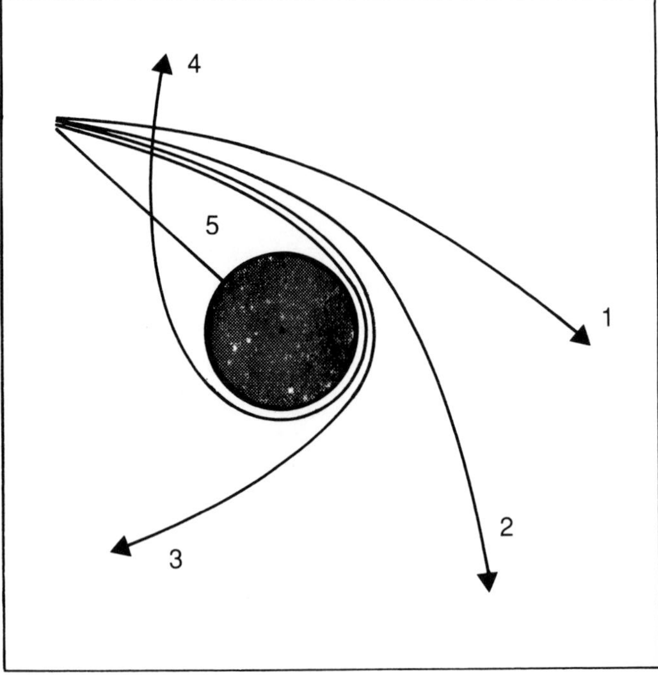

Figure 42. The Newtonian representation of gravitational forces.

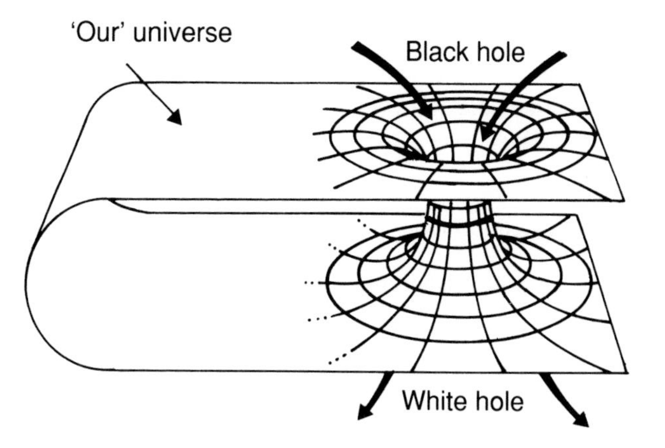

Figure 43. Identifying the sheets.

The two sheets of the Schwarzschild throat can be interpreted as two different regions of the same universe. For this the two sheets are connected at a great distance from the throat. The diagram 'cheats' by flattening the sheets, whose usual shape would be paraboloidal.

curvature of the space-time manifold and not its global shape. In particular, it allows two asymptotically flat sheets to be two different regions of the *same universe*. Mathematically this consists of cutting the sheets at a large distance from the neck and merging them to form a single surface. The operation shown in Figure 43 represents an instantaneous equatorial slice of Schwarzschild geometry.

There is still a problem. In the real universe the distances between stars, galaxies or even black holes are so great that space-time is almost always locally flat, except in the actual vicinity of a mass. Consequently the 'U-shape' caused by a distant junction of two sheets should not be curved. But it is not required: it is mathematically equivalent to represent the space-time continuum in the unfolded form of Figure 44. We now have a black hole and a white hole in the same space-time manifold at an arbitrary distance from each other, but linked by a stretched-out throat, baptised '*wormhole*' by John Wheeler.

The double nature of Schwarzschild geometry has opened the

door to the most extravagant speculations about space travel. Is it possible to travel into a black hole, through the throat and reappear through a white hole somewhere else in the universe or even in a 'parallel universe'?

Kruskal's game

To answer this interesting question, we need to know what happens *inside* the Schwarzschild throat. However, the embedding game only enables us to describe the external space-time. In particular, the singularity hidden at the heart of the black hole is not represented in the embedding. Now this singularity is a sort of joker: it is what controls the eventual free passage or otherwise to the white hole. In order to prove this, we need to play a better game, such as the one invented in 1960 by M. Kruskal.

Kruskal's technique was a very complete space-time diagram which allows us to represent on a plane the central regions of the Schwarzschild black hole. It is not straightforward to explain, but its importance and the benefits it confers make the effort worth while.

To project a two-dimensional surface onto a plane is to make a *map*. Most surfaces cannot be mapped without distortion. The most familiar example of this is a geographical map, which represents all or part of the Earth's surface on a flat plane. There are several ways of doing this. The most commonly used is Mercator's projection. The regions near the equator are accurately represented, but near the poles the representation becomes increasingly distorted. Everyone has probably noticed the exaggerated importance given to Greenland, which appears almost as large as Australia, even though in reality it is 3.5 times smaller.

Kruskal's map projects Schwarzschild space-time geometry – after the removal of two space dimensions – onto a plane by 'forcing' the light cones to remain rigid. We recall that in flat space-time without any gravitation the light cones at all events are 'parallel' to each other, their generators always inclined at 45°; but when there is a gravitational field, the cones are deformed and tilted at different angles, depending on the curvature. Kruskal's

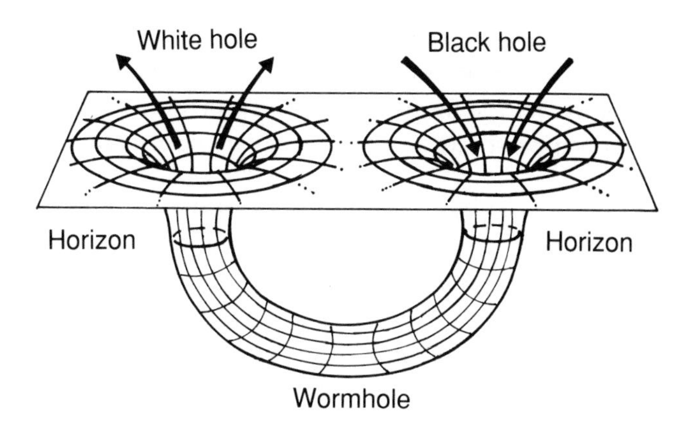

Figure 44. A wormhole in space-time.
This is the unfolded version of the preceding diagram. The horizons of the black hole and the white hole have been separated, and are linked by a wormhole.

projection requires the light cones from Schwarzschild space-time to remain parallel with each other as if there was no curvature. This type of projection is called *conformal*, a term in geometry which means that angles are preserved. To compensate for this, Kruskal's map introduces numerous distortions in space and time, but these alterations have no effect on the detailed analysis of the space-time geometry traced by the light cones.

In Kruskal's map (Figure 45) this distortion manifests itself by the fact that the trajectories of constant apparent time are straight lines which pass through the origin, while the trajectories at constant distance from the centre of the black hole become hyperbolae. The event horizon plays both roles, since it is both at a constant distance $r = 2M$ from the centre and at an infinite apparent time t. It is therefore represented by two bisecting lines in the plane, inclined at 45°, which can also be interpreted as a degenerate hyperbola reduced to its asymptotes. Furthermore, the event horizon generated by the light rays is nothing other than a light cone itself. It is therefore split into two parts: a future horizon and a past horizon.

Inside the horizon the gravitational singularity $r = 0$ also forms two arcs of a hyperbola, one in the past and the other in the future. Beyond these limiting curves Kruskal's map does not represent anything. As for the universe outside the black hole, it consists of two symmetrical sheets, one on the right of the diagram and one on the left.

How does an object move within Kruskal's map? The point of retaining a fixed network of light cones is precisely to allow us to visualise all the permitted motions, inside and outside the horizon. These trajectories must simply remain within the light cones. In other words, they cannot deviate by more than 45° from the vertical.

As an example let us consider the case of a free-falling spaceship descending into the black hole and the central singularity (curve *ABCDE*). Electromagnetic signals transmitted by the ship travel at 45° (dashed lines). A distant observer whose worldline is the arc of a hyperbola can receive only the signals transmitted at *A*, *B* and *C* with an increasing redshift as the ship approaches the horizon. The phenomenon of frozen apparent time appears quite naturally. As the ship crosses the horizon the shift becomes infinite; the light rays travel indefinitely around the event horizon, only arriving at the observer after an infinite time. The signal transmitted at *E* after the spaceship crosses the horizon cannot escape from the black hole but is condemned to be lost in the future singularity.

Kruskal's map explores the most intimate structure of Schwarzschild space-time and enables us to find unambiguous answers to questions about white holes, wormholes and passage into the 'other universe'. The region between the event horizon and the singularity is indeed a hole. But what is its colour? It is clear that the name black hole must be attached to the future horizon (upper part), where the spaceship fell in. We note on the other hand that particles and electromagnetic signals emitted in the hole at *F* can easily leave the horizon and enter the external universe. The inside of the past horizon (lower part) is therefore a *white hole*, which allows matter to fly out, the opposite of a collapse.

There still remains the question of the symmetrical sheets – right and left of the external universe. We have only to look at Kruskal's map to realise that *it is impossible to pass from one sheet in the external universe to another without going through the singularity*. In other

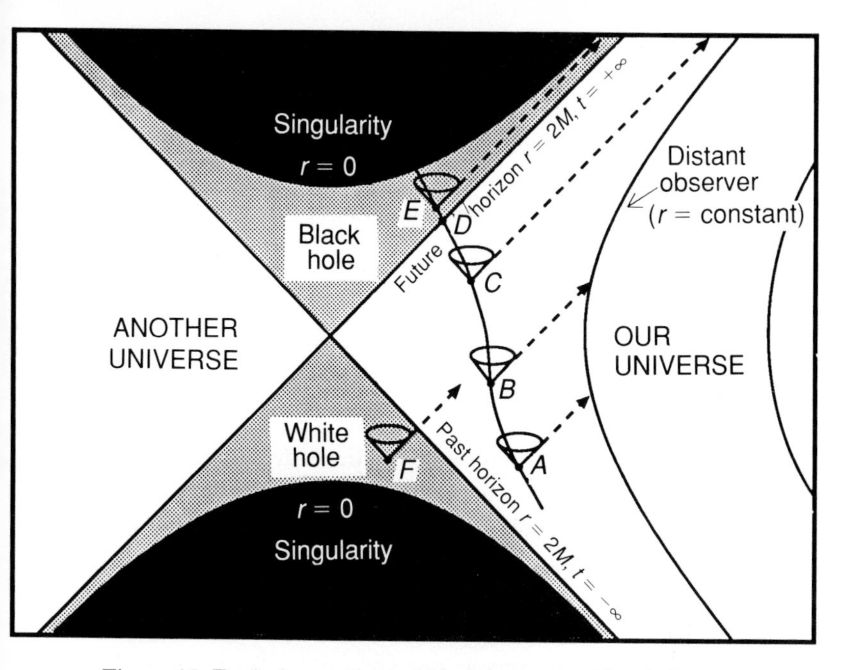

Figure 45: Exploring a spherical black hole using Kruskal's map.

The event horizon is represented by two bisectors in the plane. It is characterised by the constant distance $r = 2M$ and an infinite apparent time (future and past). The white region represents space-time outside the horizon and is divided into 'our' universe and 'another' universe. The grey zone represents space-time inside the horizon. The black regions above and below the singularity do not belong to space-time and therefore have no significance. Light rays travel along straight lines inclined at 45°, material bodies along curves which are always less than 45° from the vertical. Observers situated at a fixed distance from the black hole – on circular orbits – travel on the arc of a hyperbola which is asymptotic to the event horizon. The curve *ABCDE* is the trajectory of a spaceship falling into the black hole. It leaves its circular orbit at *A*, crosses the event horizon at *D* and reaches the singularity. The distant observer would receive the light signals from *A*, *B* and *C* with increasing delays. The light ray transmitted at *D* would reach them only after an infinite time. The light ray emitted at *E* is imprisoned in the black hole and will fall into the singularity.The light ray emitted at *F* can leave the past horizon and emerge in the outside universe, meaning that the inside of the past horizon is a white hole. It is, however, impossible to pass from our universe (on the right) to another universe (on the left) without encountering the singularity: the Schwarzschild throat is blocked!

words, the Schwarzschild throat is strangled in the middle by the infinite gravitational field of the central singularity, and nothing can pass through it.

Primordial white holes

'To be refutable is not the least charm of a theory.'

F. Nietzsche, *Beyond Good and Evil*

Anyone who has enjoyed the adventures of Alice in Wonderland will now begin to feel a little frustrated. The Schwarzschild black hole has given a shimmering vision of another universe, which it is tempting to explore, but the singularity bars the way.

Let us accept this with good grace. Often in life when things do not happen to us we console ourselves that after all they *could not* happen. Such reasoning is justified in the case of a white hole. This section will examine the *real world*, the territory, rather than the map.

The real world is too complicated. The best a physicist can do to understand observed phenomena is to develop mathematical models, which are only idealised images of what really happens. Let us consider Kruskal's map. It is an extremely powerful tool for examining space-time inside a black hole, but clearly idealised: it assumes that the gravitational source is concentrated into a point singularity which has always existed, surrounded by vacuum and hidden behind an event horizon. But in the real universe, how are black holes formed? Very probably in gravitational collapse, something very different from the symmetric situation suggested by the Kruskal map.

Let us return for a moment to the embedding surface around a non-collapsed star, drawn in Figure 39. We recall that only exterior space-time is represented by a region of Schwarzschild geometry, the rest – i.e. the interior of the star – is described by a very different geometry which depends on the structure of the stellar matter and does not contain a singularity. This leaves the suspicion that if a star collapses into a black hole *only its future history will have*

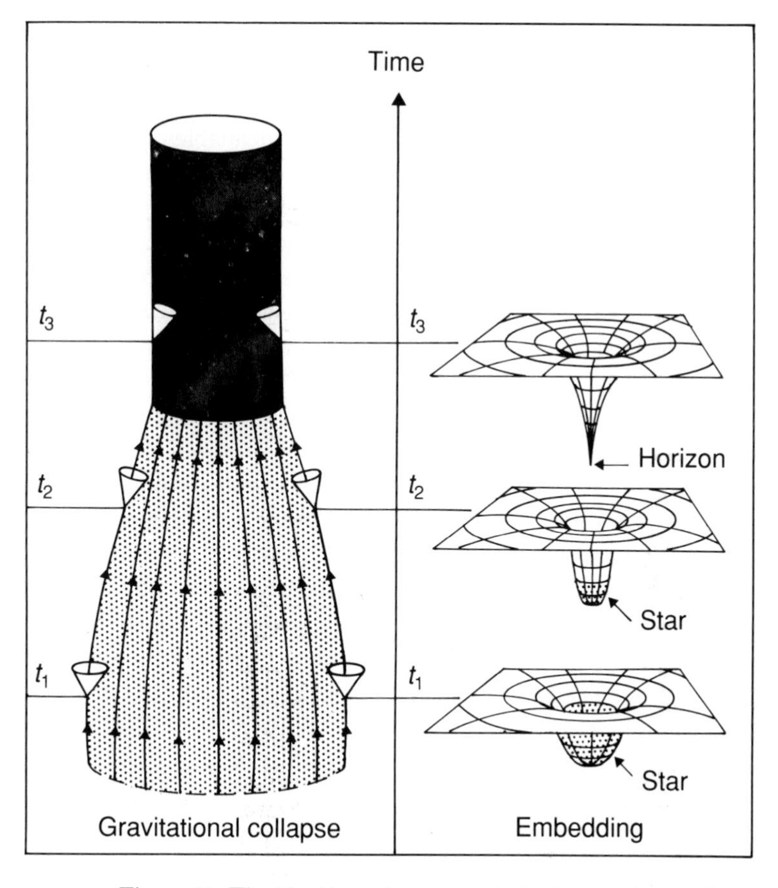

Figure 46. The blocking of a wormhole during gravitational collapse.

a physical meaning: the formation of a future event horizon and of a singularity in the future.

As the star collapses its configuration and exterior space-time change at each instant. To describe an evolving geometry, we have to re-introduce time into the embedding game. This is done in Figure 46, which shows both the evolutionary sequence of the embedding surfaces and the full space-time diagram already shown in Figure 27. The 'elastic sheet' of space-time is increasingly deformed by the collapsing star, but when the black hole is formed, the sheet does not develop a Schwarzschild throat to another universe. Instead, a kind of sharp point is formed into which the whole star disappears.

Once again the embedding technique is incapable of representing space-time inside the horizon, and the Kruskal map is needed. Figure 47 is the truncated Kruskal map showing the collapse of a spherical star. The past horizon and the past singularity have now completely disappeared, along with the symmetric sheet of the exterior universe. All that is left is a region of Schwarzschild geometry outside the horizon and a black hole and a singularity in the future.

The dual nature of a spherical black hole's space-time is nothing but a mathematical curiosity, created by the idealised symmetry of the complete Schwarzschild solution. White holes and wormholes to parallel universes cannot be formed in the real universe by the gravitational collapse of spherical stars.

It may be wondered why I have spent so long on this point. There are two reasons. First, stars are not truly spherical, and we shall see later on that rotating black holes produce myriads of wormholes that physics has not yet succeeded in blocking. Secondly, it is not impossible that there are black holes which were not formed by the gravitational collapse of a star, but which *have existed since the beginning of the Universe*, being part of the 'initial conditions'.

In general, physicists do not like to assume very special initial conditions. In this they differ from Archbishop Ussher, a contemporary of Newton, who in 1658 stated that the Universe was created as it now appears with men, animals, plants and fossils on the 23 October at 9 o'clock in the morning, 4004 BC. Modern physicists prefer to imagine a Universe born in chaos, from arbitrary conditions, with material structures appearing only later in its evolution. This argument is based on the 'Principle of Simplicity', also called 'Occam's Razor' after the fourteenth century English theologian who first formulated it. He stipulated that of a set of theories which explain the same thing, the preferred one is the simplest, requiring the least number of hypotheses. However, although the Principle of Simplicity is aesthetically seductive, it is not a logical necessity, and the hypothesis of 'pre-existing' black holes cannot be rejected at present. Only this

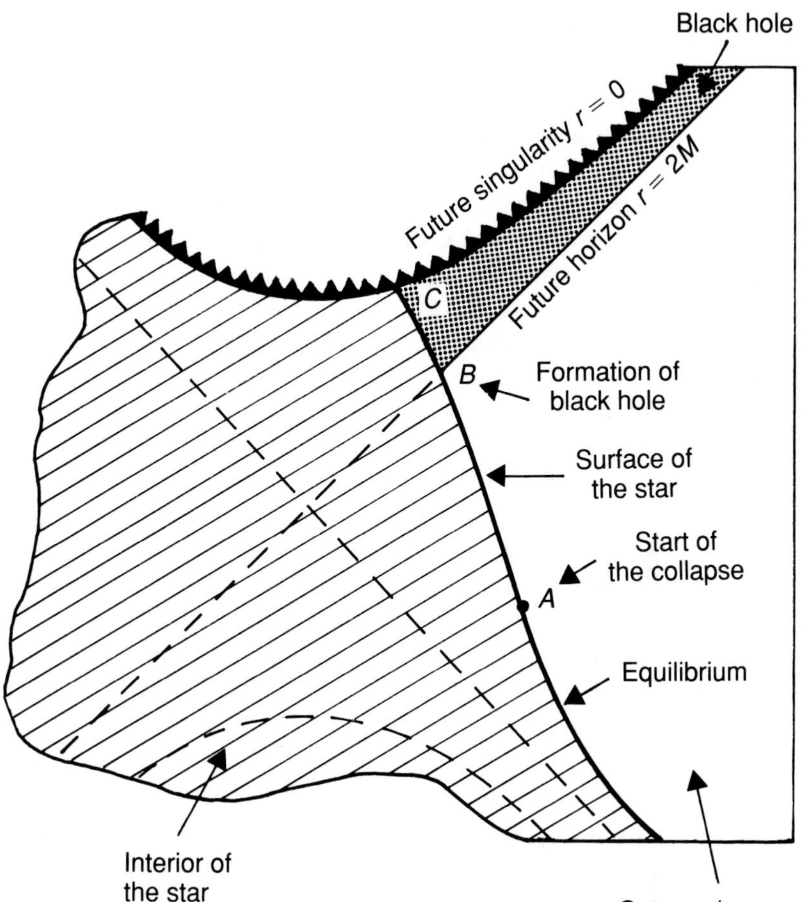

Figure 47. Truncated Kruskal map representing the collapse of a star into a black hole.

The hatched part is the interior of the star. The surface of the star is initially in equilibrium and its worldline is the arc of a hyperbola. The collapse starts at *A*. The horizon is formed at *B* and the singularity does not form until *C*. Only the future has a physical meaning. A white hole and a parallel universe do not form during the gravitational collapse of a spherical star.

type of black hole could be accompanied by its symmetric white hole.

What would a primordial white hole look like? The radiation flowing out of a white hole would be affected by two opposing effects: an 'Einstein' shift towards the red (a decrease in the frequency) because of the fact that the radiation was emanating from a strong gravitational field, and a 'Doppler' shift towards the blue (an increase in the frequency) caused by the expansion of matter leaving the hole in the direction of the observer. Some astrophysicists thought at the beginning of the 1960s that the extremely bright and distant stellar objects called *quasars*, whose nature remains mysterious, were perhaps white holes, created spontaneously from matter shortly after the 'Big Bang' in which our universe began 15 billion years ago.

Besides the objection of very special initial conditions, this model has a fatal defect: if matter was ejected from a white hole, it would collide with the surrounding matter and be slowed down to such an extent that it would collapse back on itself and form a black hole! We shall see in Part 4 that the current model of a quasar is no less fascinating. It does indeed make use of a hole, but black, and a giant.

Penrose's game

'A four year old child could understand this report. Run out and find me a four year old child. I can't make head nor tail of it.'

Groucho Marx

Our exploration of black holes is far from complete. Real black holes rotate, and we remember that their internal structure is much more complex than the static Schwarzschild black hole. To understand it better we will use a final type of map, the most sophisticated of all. Invented by the English mathematician Roger Penrose, now at the University of Oxford, it was applied to the complete description of black holes by Brandon Carter.

This game has two rules. The first is that like the Kruskal map, it is *conformal*, that is, the light cones remain straight as if there was no curvature. The second rule brings infinity in to a finite distance. Penrose's map thus allows us to represent a black hole and the entire Universe, including spatial and temporal infinity, on a single piece of paper.

We start with Minkowski's flat space-time. In this case Penrose's map is the 'diamond' of Figure 48. This is no surprise: apart from the requirement of travelling within 45° of the vertical, nothing perturbs the trajectories of matter and radiation. The Universe of Special Relativity, empty of gravity, is just a uniformly flat desert.

Let us now consider the map of the static Schwarzschild black hole. Penrose's map (Figure 49) is little different from Kruskal's, except that space-time now has boundaries at a finite distance. It clearly shows that the Schwarzschild singularity – divided into a past singularity and a future singularity – is a *boundary of space-time*, in the same way as infinite distance. This singularity is a *horizontal line* which no trajectory entering the black hole can avoid. It is *space-like* (i.e. parallel to the space axis) and marks the proper *end of time* for all explorers of the black hole.[5]

Apart from the properties of the singularity, Penrose's map makes it easy to see the fundamental characteristics of space-time inside a black hole, which are by now familiar: the double sheet structure of the regions outside and inside the event horizon, the interchange of the space and time directions making it impossible to maintain a fixed position inside a black hole and the impossibility of passing from one exterior space-time to the other through the Schwarzschild throat.

Strait is the gate

Penrose's mapping technique comes into its own when it is applied to the Kerr space-time generated by a rotating black hole. The resulting diagram is much more complicated than that of a static black hole (Figure 50). It consists of blocks which

[5] It should be noted that the past singularity does not represent any danger for the astronaut, for in order to encounter it he would have to travel backwards in time.

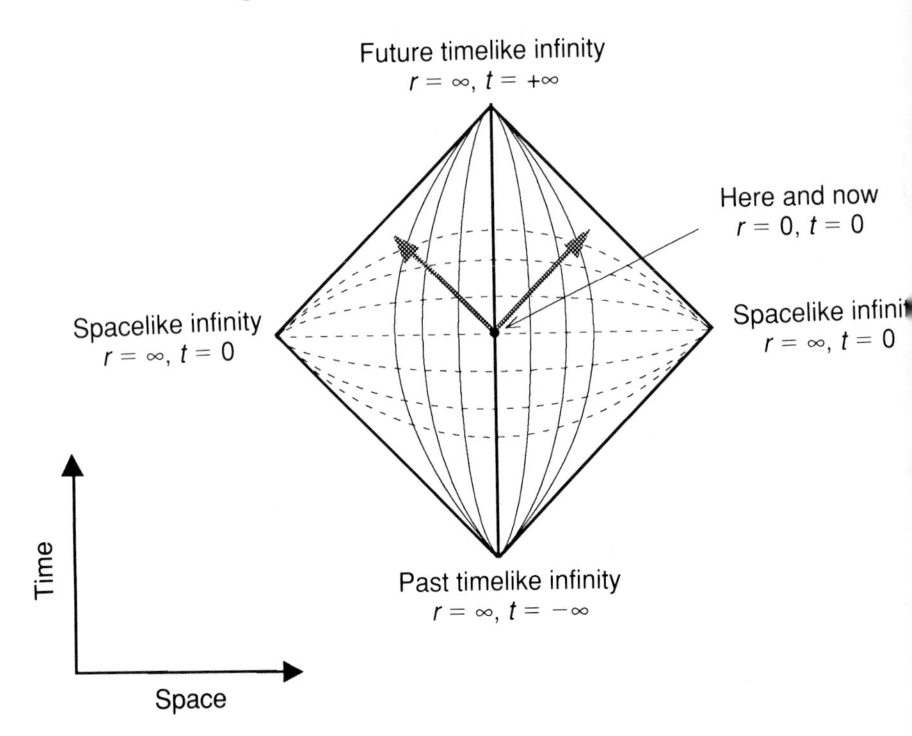

Figure 48. The Minkowski diamond.

As in all space-time diagrams, space r is measured horizontally and time t vertically. The 'here and now' position is at the centre of the diamond ($r = 0$, $t = 0$). 'Present' spatial infinity ($r = \infty$, $t = 0$) is represented by the left- and right-hand vertices of the diamond. Time-like infinity ($t = -\infty$, $t = +\infty$) is represented by the lower and upper vertices of the diamond (past and future). All the permitted worldlines of matter and radiation flow without incident from the lower to the upper edges. The light rays travel along the lines inclined at 45°, matter along curves inclined at less than 45° to the vertical. In particular, the trajectories of constant position r join the lower and upper vertices of the diamond. The curves of constant time t (broken lines) join the left- and right-hand vertices of the diamond. These curves cross the light cones; they are of course forbidden trajectories because of the impossibility of evolving at constant time.

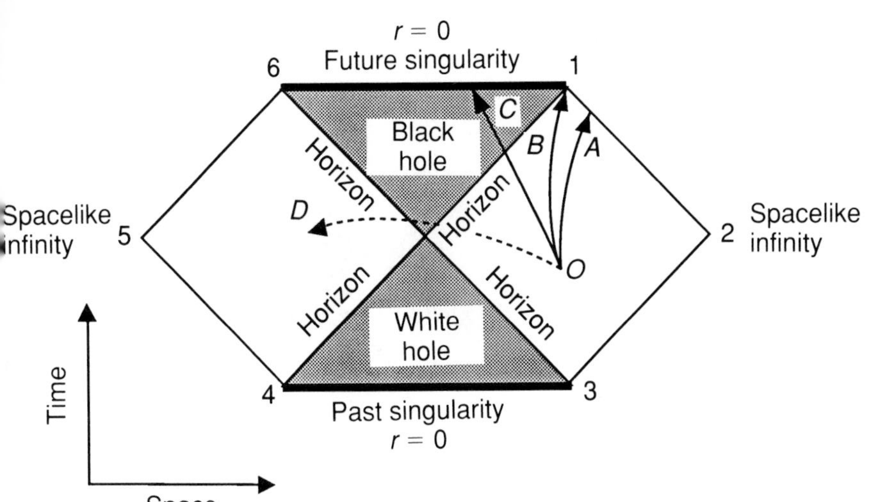

Figure 49: Penrose map of a static black hole.

The complete Schwarzschild black hole space-time is contained in a hexagon, whose vertices are numbered 1 to 6. The double structure of the exterior space-time is recognisable (the squares, right and left, bounded by the horizon and null infinity) and the interior universe of the black hole (upper and lower triangles, bounded by the horizon and the singularity). Several possible journeys are drawn, from an initial position *O*. The curve *A* is followed by an astronaut travelling away from a black hole with constant acceleration, his velocity tending towards that of light. Curve *B* is followed by an astronaut who remains at a constant distance from the black hole, it necessarily ends at vertex 1. Curve *C* is that of an astronaut exploring the interior of the black hole. Once he has crossed the event horizon, there is a fundamental change in the space-time structure. Outside the black hole the curves characterised by a constant distance from a black hole have to go from vertex 3 to vertex 1 (or vertex 4 to vertex 6 in the symmetrical sheet), but inside the hole curves of constant position join 1 and 6 or 3 and 4. These curves lie outside the light cones and are thus forbidden. We recall the impossibility of maintaining a fixed position inside the black hole, just as it is impossible to evolve at a constant time in the exterior universe. Finally, the trajectory *D* (broken line) is the passage from one sheet of the exterior universe to the other through the Schwarzschild throat, avoiding the central singularity. It is forbidden since it requires a velocity greater than that of light. The Schwarzschild throat is blocked by the singularity.

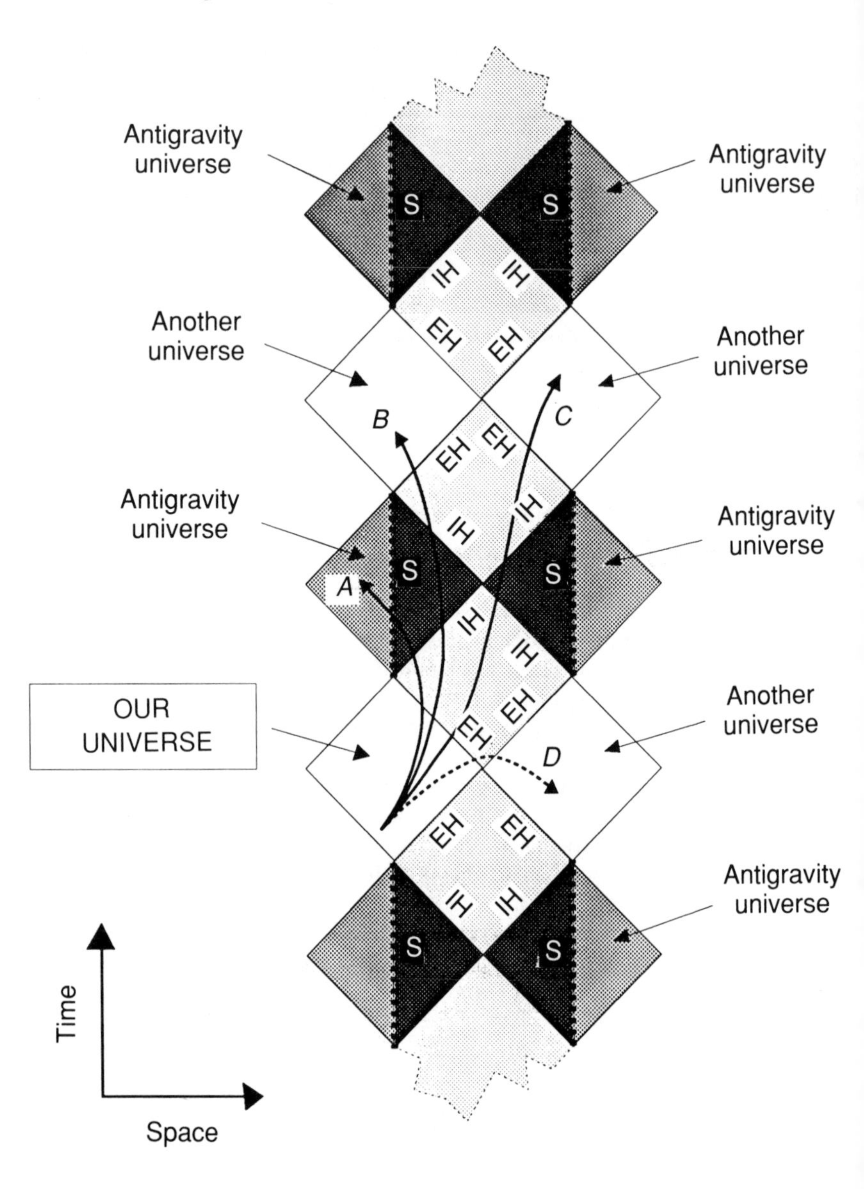

Figure 50 : Penrose map of a rotating black hole.

The diagram repeats indefinitely to the past and future. The universes outside the black hole are the white squares, the universes inside the black hole are the light and dark shaded squares. Null infinity is written as ∞, the external event horizon

repeat indefinitely into the past and future, with an *infinity of exterior universes* to the black hole and an *infinity of interior universes*!

The exterior universes are bordered by null infinities and the event horizon. The interior universes each contain a singularity dividing them into several regions. The rotating black hole has an inner event horizon which surrounds the central singularity. Each time a horizon is crossed the direction of space and time are exchanged. In travelling from the exterior universe to the singularity, an astronaut is subjected to two consecutive changes. Consequently, within the inner event horizon, the directions of space and time are exactly the same as outside the black hole.

This is the reason why the singularity is *vertical*, not horizontal. In fact, it is no longer even a boundary of space-time. There exists a region situated *on the other side* of the singularity. This is easy to understand if you remember that the singularity is not reduced to a central point at $r = 0$ as in a static hole, but a ring lying in the

Caption for Fig. 50 (*cont.*)

– the true boundary of the black hole – is written as EH, the inner event horizon is written as IH, and the singularity is called S. The interior universes are split into several regions. The directions of space and time are exchanged in the light shaded region between the horizons EH and IH: it is impossible to maintain a fixed position. In the dark zone between IH and S the directions of space and time are the same as those in the exterior universes. The dotted region 'on the other side' of the singularity stretches to an infinite distance, but the distances are 'negative'. This is the antigravity universe. The singularity is oriented vertically and is discontinuous, meaning that it can be avoided and it is possible to pass through it. In travelling within a rotating black hole, the only rule is to remain within 45° of the vertical. Several trajectories starting in the exterior (our) universe are shown. *A* passes through the singularity and explores the antigravity universe. *B* and *C* pass through four event horizons and reappear in another universe. *D* is forbidden because it requires a velocity faster than that of light. It is mathematically possible to identify the 'other' universes as our universe, but the operation gives rise to time paradoxes.

equatorial plane.[6] This ring does not define an edge to the space-time geometry because the astronaut can pass through it. The ring singularity of a rotating black hole thus lacks the inexorable character of the point singularity. It is not a space-like but a *time-like* singularity (i.e. parallel to the time axis), and because of this it does not represent the end of time for explorers of black holes.[7] Apart from the dangers of tidal forces, a traveller could come within a hair's breadth of the singularity provided he does not touch it. He could even see if light signals came from it.

As to the 'other side' of the singularity, this is a region of a spatially infinite space-time in which the distances are 'negative'. This apparent absurdity is interpreted as an inversion in the attractive character of gravitation. It would become *repulsive*, forcing matter to distance itself indefinitely from the singularity.

The rich structure of the rotating black hole provides fascinating possibilities for exploration. In Figure 50, trajectory *A* shows the possibility of exploring the antigravity universe on the other side of the singularity. Trajectories *B* and *C* show that it is theoretically possible to penetrate the interior of a black hole – preferably a very large one, so that the spaceship is not destroyed by tidal forces – fly through the singularity and emerge into other exterior universes. Finally, trajectory *D* is forbidden, because it leaves the light cone. So even with this map there is still one part of space-time whose exploration is strictly forbidden.

The time machine

'Common sense, however it tries, cannot avoid being surprised from time to time. The aim of science is to save it from such surprises.'

Bertrand Russell

The Penrose map becomes quite worrying for our mental health when we try to interpret the exterior universes as sheets of a

[6] See Figure 37.

[7] Except for those astronauts unwise enough to travel exactly in the equatorial plane!

single universe, although this is perfectly acceptable in General Relativity. In this case the rotating black hole would connect myriads of wormholes to different parts of the space-time geometry. Since two events can differ in time as well as in space, it would be possible, in principle at least, to pass from one given position at a given time through a carefully chosen wormhole and arrive at the same position, but at a *different time* in the past or future. In other words, the black hole could be a sort of *time travel machine*.

The history of science has shown many times that what seems absurd one day can become an accepted fact the next. None the less, a journey back through time is an affront to common sense. It is difficult to accept that a man could travel back through time and kill his grandfather even before he has had the time to produce children. For the murderer could not have been born, and could not have murdered his grandfather, who could then have had descendants, who could have murdered him, and so on . . . This time paradox was related by the French writer René Barjavel, in a story called *Le voyageur imprudent*.

A journey into the past violates the law of causality, which requires that the cause always precedes the effect.[8] However, causality is a rule imposed by logic, and not by the theory of relativity. Causality is implicit in Special Relativity, where there is no gravitation. Here, travelling into the past requires motion faster than the velocity of light, and is absolutely forbidden. However, in General Relativity the Universe is curved by gravitation, and the space-time geometry can be distorted – by a rotating black hole, for example – enabling the past to be explored without having to go faster than the velocity of light.

If a journey into the past is possible, is common sense irretrievably lost? Not necessarily, if we replace the principle of causality by that of consistency: the new rule would then stipulate that the evolution of a physical system must be self-consistent, even if travel into the past is allowed. The situation described by Barjavel (murder of one's forbears) is clearly not consistent. But

[8] See page 23.

theoreticians who enjoy puzzles have imagined acausal situations so complicated that common sense is baffled.

To avoid this, could we save matters by arguing that the potentially acausal distortions caused by a black hole are just mathematical artefacts and do not really occur when a rotating star collapses to form a black hole? In the spherical case it is possible to follow the evolution of the exterior and interior geometry of the contracting star step by step, and to use a truncated Kruskal map or a set of embedding diagrams to see that the disconcerting phenomena of the white hole and anti-universe are naturally eliminated. Unfortunately, in the non-spherical case we do not know how to map rigorously the exterior or interior space-time geometry of a rotating star. It is continuously perturbed by gravitational waves, and only when the black hole has formed is the Kerr geometry enforced. Recent calculations have shown that any matter or radiation entering a rotating black hole has its energy so amplified by the gravitational field that its self-gravity would change the space-time and block the wormhole. Recently theoreticians have asked themselves under what conditions a macroscopic wormhole (for example associated with a giant black hole, so that the tidal forces are not too large) could remain open despite the entry of matter (for example a spaceship). I remember in 1976, the year that I began research in General Relativity, a serious British foundation (The Bacon Foundation) offered a prize of £300 for the solution of the following problem: 'according to current theory, rotating black holes are the actual gateways to other regions in space-time. How therefore could a space vehicle pass through a rotating black hole into another region of space-time without being crushed by the gravitational field of a singularity?' As a beginner in the subject I was obviously unable to win the prize, or even to attempt it. I do not know if anyone won the prize, but it is clear that the problem was not solved until 1985, and in a rather exotic way: it turns out that only those wormholes that are laced with matter exerting an enormous 'negative pressure' can be stable. A negative pressure is a tension, like that of a spring that one stretches. In ordinary matter tensions are always much smaller than the energy (the breaking tension of steel is, for example, 1000 billion times smaller than its energy content per unit volume). In the exotic matter required to stabilise a wormhole, the proportions would be inverted. All of this is clearly highly speculative; no-one has the faintest idea if this type of 'negative' matter can exist in nature. To use these short cuts in space-time effectively we would have to construct 'negative' wormholes, perhaps by growing a microscopic negative wormhole. Even allowing this fantastic idea, there is nothing telling us that a spaceship made of normal matter could cross this negative energy region safely. This vague theoretical-artistic possibility enabled the popular American astronomer-writer Carl Sagan to construct his novel *Contact*, using the idea of communication with extra-terrestrial civilisations by means of wormholes. But although this is an exciting story, it is pure science fiction, and very likely to remain so!

Gravitational singularities

The potential violation of causality does not endanger the actual theory of black holes, but brings into question the true nature of the singularity and the 'fine' structure of space-time. Here we reach the frontiers of contemporary physics.

We may ask ourselves first if the appearance of a singularity, indefinitely crushing matter and the space-time geometry, is not the result of a naive application of General Relativity to the problem of gravitational collapse. Singularities appear in the more general background of *cosmology*, the branch of astrophysics which deals with the evolution of the Universe as a whole. In the Big Bang theory, the Universe was born in a singularity about 15 billion years ago. This is strongly supported by observations of the expansion of the Universe and the cosmic microwave radiation, the cold remains of its birth. However, in cosmology as elsewhere, the models used to describe the past and the present state of the Universe are very idealised. Here too it is legitimate to wonder if the cosmic singularity is not just an unwanted by-product of mathematical simplification.

Two English scientists, Stephen Hawking from the University of Cambridge, and Roger Penrose, who invented the conformal map, demonstrated in the 1960s that this is not the case. Singularities form an integral part of General Relativity. We may not be able to demonstrate that the gravitational collapse of a 'real' star leads to the formation of an event horizon and a black hole, but we can prove that it will inevitably result in a singularity. Hawking and Penrose also established that if we extrapolate backwards into the cosmic past, all models of the Universe which actually agree with what is observed now must have begun with a singularity. If the Universe contains sufficient matter, it may even end in a singularity; the expansion phase will eventually give way to a symmetrical contraction phase, a truly universal collapse.

These very important theorems generalise a result already known from the Newtonian theory of gravity: a cloud of dust particles contracts under its mutual gravitational attraction to an infinitely dense singularity. Thus, singularities turn out to be an unavoidable consequence of the attractive and 'self-accelerating' properties of gravity. How do we face up to them?

Cosmic censorship

'Nature likes to hide herself.'

Heraclitus (500 BC)

The gravitational collapse of a star to a singularity might occur in one of two ways, depending on whether a black hole is formed or not. If a black hole is formed, the event horizon hides everything within it, including the final crushing of matter into the singularity. This happens in spherical collapse and it matters little to a physicist living in the exterior universe to know whether a singularity forms or not. Since the interior of the black hole cannot communicate with the exterior, the laws of nature and common sense could be violated near the singularity without the world of physicists knowing anything about it.

In the second possibility, the singularity forms without a black hole to hide it. Let us imagine, for example, that a rapidly rotating massive star retains an angular momentum greater than the critical value during its collapse. In this case the formation of a stable horizon of a black hole is rendered impossible by centrifugal forces and the singularity is left *naked*. Particles or electromagnetic signals could escape from it and be observed at large distances. Now because of the infinities associated with a singularity, its effect on the space-time geometry would be totally unpredictable. Without the protection of the event horizon, physicists would be ripe for unemployment, because all calculations and predictions made one day could be contradicted the next by the whim of a naked singularity!

Obviously, a naked singularity has never been observed in the Universe. However, this does not prove that they do not exist. To avoid this embarrassing situation Roger Penrose formed an hypothesis according to which Nature forbids the existence of naked singularities. According to the hypothesis, gravitational collapse always clothes the singularity in an event horizon. This conjecture is called *Cosmic Censorship*.

The idea of Cosmic Censorship is reassuring, but it has never been rigorously proved within General Relativity. The conjecture works in situations which do not differ much from the spherical case. For more extreme situations, the question remains completely open. There is another cause for worry: the cosmic singularity which is supposed to have given birth to our Universe 15 billion years ago, was not hidden behind an event horizon.

Quantum gravitation

'If the Almighty had consulted me before the creation of the World, I would have recommended something simpler!'

Alphonse X the Wise (thirteenth century)

Even if the conjecture of Cosmic Censorship could be rigorously proved, it would still not resolve the problems of gravitational 'anomalies'. The ring singularity, although hidden in a rotating black hole, allows objects to pass through wormholes and is thus implicated in the violation of causality.

The real problem is therefore not to know if naked singularities offend or not but, whether they exist in the real world. To discover this we have to return to the source of the problem: General

Relativity. How can a theory which predicts configurations where certain physical quantities become infinite be correct?

Science has often produced theories with singularities which have then been eliminated by the arrival of improved theories. A good example of this is the early model of the atom, regarded as a miniature planetary system controlled by electric forces. In the theory developed at the beginning of the century by Ernest Rutherford, electrons orbiting the atomic core ought to have rapidly lost their energy and plunged into the nucleus. But our experience shows that atoms are stable. The abnormal behaviour of Rutherford's atom showed therefore that the theory was incomplete. The development of quantum physics resolved the problem. In this new theory the energy levels of the electrons were quantised, stabilising the model atom and removing the singularity.

It is tempting to make the analogy with General Relativity. The occurrence of gravitational singularities demonstrated by Hawking and Penrose probably shows that the theory is being applied outside its domain of validity. Could quantum physics remedy this?

The first part of the answer is suggested by a closer examination of Hawking's and Penrose's theories. Their conclusion depends on the seemingly reasonable hypothesis that 'matter has positive energy'. This condition is apparently satisfied by all known forms of matter, including extreme forms such as neutron stars, which although not reproducible in the laboratory can be extrapolated from our knowledge of nuclear matter. However, even if all 'classical' matter has a positive energy, this is not true of *quantum* matter. Recent calculations have shown that certain phenomena in elementary particle physics violate the condition of positive energy.[9]

This is the crux of the problem. Although General Relativity is the best theory of gravity we have, it is obviously incomplete because it does not take account of the principles of quantum mechanics which govern the evolution of the microscopic world. Now the singularity phenomenon involves precisely the structure of space-time on a very small scale. It is hardly surprising then that the

[9] For example the spontaneous creation of particles in a vacuum, made possible by quantum mechanics, (see Chapter 14).

application of a classical theory to a quantum domain produces these undesirable singularities.

The actual relation between quantum mechanics and General Relativity is even more distant. The first governs the domain of elementary particles, which move under the action of nuclear forces which have very short range. Its main characteristic is that it gives a 'fuzzy' description of phenomena, in which events can be calculated only in terms of probabilities. Electromagnetic forces govern the transitional region, including human beings. In some phenomena (lasers, transistors and so on) quantum mechanics plays a crucial role, but in others (propagation of radio waves and the like) its role is negligible. Finally, on the astronomical scale, quantum effects become totally blurred and 'classical' gravitation as described by General Relativity takes over.

But, in the words of Sheldon Glashow,[10] it might well be that 'the snake is eating its own tail'. Some physicists believe that gravity becomes the dominant force at scales smaller than 10^{-33} centimetre. This minute length was introduced in another context a century earlier by Max Planck. It is obtained by a clever combination of the fundamental constants of Nature (the gravitational constant, velocity of light and Planck's constant) and is independent of the properties of elementary particles. It represents the smallest scale above which the space-time geometry can still be considered as smooth. Below this scale the texture of space-time itself may not even be continuous, but like all energy and matter, consist of small grains. In the formulation by John Wheeler, this is where 'the fiery marriage between General Relativity and quantum mechanics' will be consumated. The offspring will obviously be called *quantum gravity*.

I use the future tense, because quantum gravity is still more an idea than a theory. During the last 40 years of his life, Einstein tried in vain to unify General Relativity and quantum mechanics. Today hundreds of theorists are working on this enormous problem. Apart from the discouraging mathematical difficulties there is a total lack of concrete experimental data. Its domain of application – in distances or energies – is fantastically removed from the laboratory.

[10] Nobel prize for Physics in 1979.

It is possible to use the big particle accelerators to probe distances comparable to the radius of an elementary particle such as a proton, at 10^{-13} centimetre,[11] but the gulf between us and quantum space-time is immense: the ratio between the radius of a proton and the Planck length is of the same order as the ratio between the size of our Galaxy and of a human being.

Fortunately for contemporary physics, new ideas abound despite these unfavourable conditions. John Wheeler has suggested that the geometry of microscopic space-time is turbulent and continuously changing, agitated by quantum fluctuations. It can be compared with the surface of an ocean (Figure 51). Seen from an aeroplane the ocean appears smooth. At a lower altitude the surface remains continuous but is fluctuating. Closer still it is very turbulent, even discontinuous, because as the waves break droplets of water are thrown into the air. In the same way, while space-time structure appears continuous at our level, its 'foam' may be apparent at the scale of the Planck length and could produce droplets which would appear to us as elementary particles.

The most recent attempts to explain this idea called for the help of a 'superspace', in which the number of dimensions would be more than four.[12] In our everyday life only three dimensions of space and one of time are perceptible, but the real Universe may close on itself in additional dimensions whose characteristic length would be the Planck length. A simple illustration is that of a long piece of flexible piping. It has two dimensions: one is slightly curved in the direction of its length and the other is very curved and much smaller in the direction of the circumference. Now seen from a distance the piping would just appear as a piece of thread with just one dimension and no curvature.

In spite of these fascinating speculations, no suitable scheme has yet emerged. Because of the lack of experimental verification, physicists have to rely on theoretical requirements. One of these is precisely the elimination of gravitational singularities. They would be replaced by quantum fluctuations of the space-time geometry,

[11] The highest energies actually obtained in particle accelerators enable us to probe the properties of matter down to 10^{-18} metres.

[12] It could even be that the true dimension of 'superspace is 'fractal', that is, not an integer!

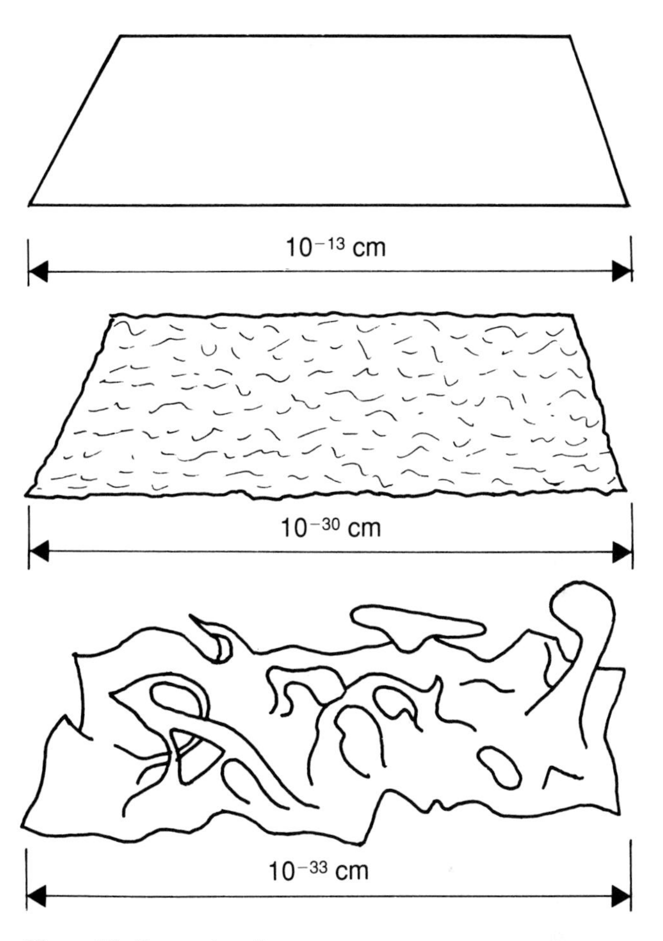

Figure 51: Space-time foam.

Seen from a distance space-time is smooth. At the scale of the Planck length, it is very complicated and continuously changing.

which would not give rise to infinite physical quantities but would have the effect of blocking the wormholes of rotating black holes. This is the price of safeguarding causality.

What is certain is that black holes have a key role to play in the development of quantum gravity. Recently some workers have proposed a model in which microscopic wormholes (with a scale 100 billion billion times smaller than an atomic nucleus), through their contribution to the quantum mechanics of space-time, help to fix the values of all the fundamental constants of Nature. Born two centuries ago amid indifference, and only now grown up, black holes are really just beginning their career. The next two chapters show in more detail how black holes reveal a profound connection between two apparently separate domains of physics.

13

The black hole machine

Thermodynamics

Physicists have always tried to understand why the Universe is ordered and not chaotic. From distant galaxies to the living cell, the Universe has developed complex structures of all scales. The diversity and complexity of organised systems is so great that it would seem presumptuous to try to understand the general principles which govern the organisation of systems as different as human beings and stars. This, however, is in part achieved by *thermodynamics*.

This discipline began in the nineteenth century. Its original aim was prosaic enough; engineers and industrial companies wanted to control heat exchanges and mechanical energy so as to make steam engines more efficient. From this small beginning profound universal concepts were developed which can be applied to the evolution of most physical systems.[1]

The strength of thermodynamics is that it sets up very general laws which do not depend on the detailed structure of systems. For example, it enables us to understand the thermal properties of materials without knowing their atomic structure. Thermodynamics can be summed up in four laws, numbered zero to three for historical reasons.

Law 0 states that all parts of a system in thermal equilibrium have the same temperature. Law 1 states that heat is a form of energy

[1] The strict application of thermodynamics to the 'living' world raises difficulties.

and describes how different forms of energy are exchanged in an evolving system. If two systems with different temperatures come into contact – for example a litre of hot water and a litre of cold water – heat exchange takes place between the two systems until an equilibrium temperature is reached. This will of course be somewhere between the two initial temperatures, giving two litres of lukewarm water. Law 3 asserts the inaccessibility of absolute zero (-273.15 °C). The temperature of a system can be decreased by appropriate transformations, but it is impossible to cool it *completely* by a finite number of transformations.[2] Law 2, however, is the centre-piece of thermodynamics, because it has the most universal field of application. In simple terms, it asserts that systems become more and more disordered as they evolve. Mountains are eroded, houses collapse, cars break down, stars explode and men grow old and die. Obviously, more ordered structures are being formed all the time: births, crystal growth, construction of cities, and so on. However, the creation of order in a part of a system has to be paid for by an increase in disorder in the full system. Physicists have invented a quantity to measure disorder: it is called *entropy*. The precise form of the second law of thermodynamics states that the entropy in an isolated system can only increase with time.

Entropy has a real significance: it measures *disorder*. Let us see why. Mathematically, entropy calculates the total number of internal configurations a system can adopt without changing its external state. For example, the external state of a gas is fixed by its temperature and pressure, but there are a tremendous number of possible disordered motions of the gas molecules, all corresponding to the same temperature and pressure. It is this very large number which fixes the entropy of the gas. In the same way, the 'macroscopic' state of a sugar cube is determined by several overall parameters, such as chemical composition, temperature and volume, but each macroscopic state corresponds to an enormous number of hidden microscopic states, depending in particular on the molecular structure and internal vibrations. The entropy of a system measures the number of hidden internal configurations and

[2] In the laboratory it is currently possible to reduce the temperature of helium 3 to 10^{-6} K by the use of cooling lasers.

therefore is a measure of our ignorance of the details of a system. The more organised a system is, the lower its entropy, and vice versa. It is a measure of disorder.

The notion of entropy can be made even more general by relating it to the notion of *information*. It is clear that the microscopic configurations of a system contain hidden information about the system. The more information that is hidden the greater the entropy, and at the same time the amount of available information is reduced. A very ordered system has a lot of available information and therefore a low entropy. By ordering the letters in this book, I have communicated a lot of information to my readers – that was my aim, at least! But if I suddenly chose to arrange the letters haphazardly, the information content of the book would become practically zero, save for the fact that there was an author. In other words, *entropy measures the lack of information on a system.*

The dynamics of black holes

A black hole is not a passive body, jealously hiding a mass destined to remain inert forever. Because of its electric charge, and above all its angular momentum, a black hole is a *dynamical system*, capable of being subjected to and exerting forces, absorbing or supplying energy. In other words it changes with time. It is therefore important to examine the laws governing the evolution of black holes and compare them with the laws of thermodynamics.

Usually in thermodynamics the state of a system can be characterised by two fundamental parameters: its temperature and its entropy. The laws of thermodynamics state precisely how the other macroscopic parameters, such as energy, volume or pressure, vary as a function of temperature and entropy during a transformation of the system. In the same way, the dynamic state of a black hole is characterised by two parameters: the *area of the black hole*, which measures the surface of the event horizon, and the *surface gravity* which measures the acceleration due to gravity at the horizon.

Since the black hole *equilibrium state* depends on only three parameters, mass, angular momentum and charge, the area and the surface gravity of the black hole can also be expressed as a function of these three parameters. For Schwarzschild's static black hole, characterised uniquely by mass, these calculations are particularly simple. The event horizon is a sphere whose radius is proportional to the mass of the black hole ($r = 2M$); its area is therefore proportional to the square of the mass. A 10 $M_\odot$ spherical black hole has an area of 5650 square kilometres, comparable to the size of a county. Similarly, the surface gravity is inversely proportional to the mass. A 10 $M_\odot$ spherical black hole has a surface gravity 150 billion times that of the Earth.

The dynamics of black holes can be summed up in four laws, bearing a striking resemblance to the usual laws of thermodynamics.

Law 0 states that all the points on the event horizon of a black hole at equilibrium have the same surface gravity. This is surprising when you think of the flattening of the poles caused by the centrifugal force. Now for ordinary rotating celestial bodies such as the Earth, we know that gravity is higher at the poles than at the equator. By contrast, no matter how flat the event horizon is, there is no variation in the surface gravity of a black hole.

Law 1 states how, during a transformation of a black hole (resulting from the capture of a dust cloud or an asteroid, for example), its mass, rotational velocity and angular momentum vary as a function of its area and surface gravity.

Law 3 states that it is impossible to reduce the surface gravity of a black hole to zero by a finite number of transformations. An example of a black hole with zero surface gravity is Kerr's maximal black hole, whose angular momentum has reached the critical value. The third law states that a maximal black hole is a limit which cannot be reached in nature. For a slowly rotating black hole, it is possible to increase its angular momentum by dropping pieces of matter into it from appropriate orbits, but the maximal state will remain inaccessible.

Finally, Law 2 of black hole dynamics asserts that the area of a black hole can never decrease with time. If a completely isolated black hole maintains a constant entropy, it will increase its surface area when matter or radiation is captured. In the same way, if two black holes collide they form a single black hole whose area is greater than the sum of the two individual areas (Figure 52).

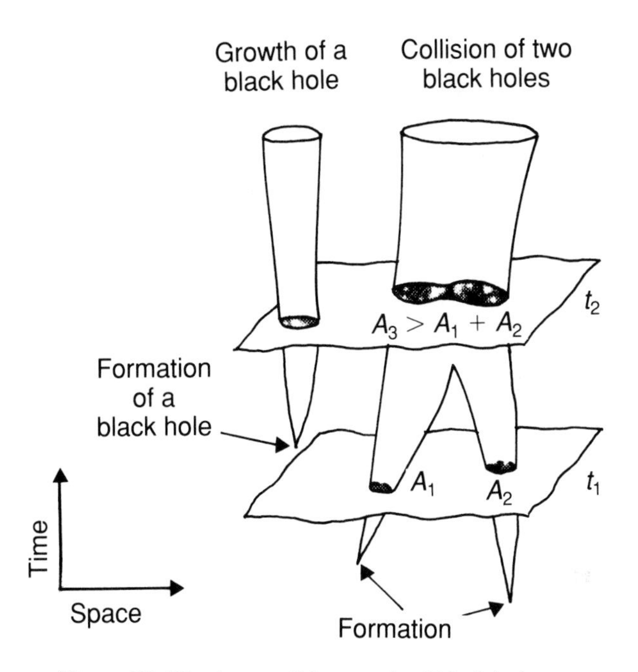

Figure 52. The irreversible growth of black holes.

During its evolution a black hole can only increase its surface area. If two black holes collide they form a single black hole whose surface A_3 is greater than the sum of the surfaces A_1 and A_2 of the parent black holes.

This basic result, discovered by Stephen Hawking, shows the close connection between the area of a black hole and the entropy of a thermodynamic system. Can we take this analogy a step further and say that a black hole actually possesses entropy?

The Israeli physicist Jacob Bekenstein thought so. A black hole is a cosmic prison preventing all matter and radiation – and therefore all information – from escaping. On the other hand, when a material body disappears into a black hole, all knowledge about its internal properties is lost to the external observer; all that is left is the black hole's new value of mass, angular momentum and charge. Consequently the black hole *swallows all information.* It must therefore have an entropy. As in thermodynamics, this measures the total number of possible internal configurations corresponding to a given state. The calculation gives a result which is in fact proportional to the area of the black hole.

The entropy of a 1 $M_\odot$ black hole is a billion times greater than the Sun's. The difference can be explained by the fact that, during its formation the black hole 'lost its hair', that is, it swallowed all the information about the matter other than the mass, charge and angular momentum. It is for this reason that black holes are the greatest reservoirs of entropy in the Universe.

The black hole as a source of energy

The first law of black hole dynamics says that although a black hole prevents any radiation or matter from escaping, it can give up energy to the external medium. In fact, the total mass-energy of a black hole can be split into three components: a 'rotational' energy associated with the angular momentum, an 'electric' energy associated with the charge and an 'inert' mass-energy. The Greek physicist Demetrios Christodolou proved that the first two types of energy could be extracted from the black hole, but the third remains irreducible. This irreducible energy is directly linked to the area of the black hole, which according to the second law cannot be reduced during a transformation.[3]

The spherical and neutral Schwarzschild black hole is the one with minimum energy. It remains a gravitational well, swallowing particles and radiation and increasing its mass with each interaction. Conversely a black hole near to its maximal state is full of energy and is not mean with it. Its rotational energy, representing at least a third of its total energy, can be extracted.

The amount of energy which can potentially be extracted is a fantastic quantity, in comparison with which the explosion of a supernova appears like a damp squib. However, the extraction of rotational energy from a black hole does not have the cataclysmic character of a stellar explosion. It can be achieved only with extreme care. The key role is played by the ergosphere, the region between the static limit and the event horizon. Roger Penrose suggested the following extraction mechanism.

[3] The entropy can at best remain constant, in reversible transformations.

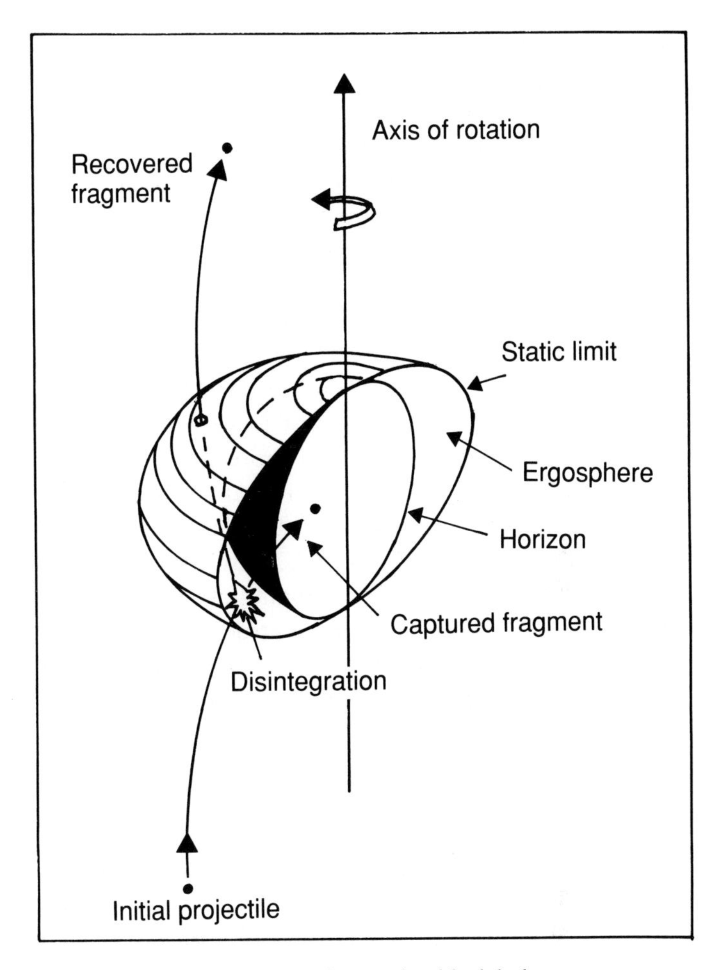

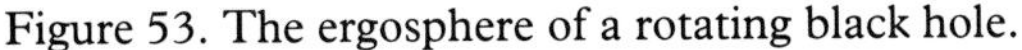
Figure 53. The ergosphere of a rotating black hole.

When a projectile disintegrates in the ergosphere and one of the fragments falls into the black hole, the other fragment can leave and be recovered, carrying more energy than the initial projectile. (After Ruffini and Wheeler.)

A distant experimenter fires a projectile in the direction of the ergosphere. When it arrives the projectile splits into two pieces: one of the pieces is captured by the black hole, while the other flies out of the ergosphere and is recovered by the experimenter (Figure 53). Penrose demonstrated that the experimenter could direct the projectile in such a way that the returning piece had a greater energy than that of the initial projectile. This is possible if the fragment captured by the black hole is travelling in a retrograde orbit (that is orbiting in the opposite sense to the rotation of the black hole), so that when it penetrates the black hole it slightly reduces the hole's angular momentum. The net result is that the black hole loses some of its rotational energy and the difference is carried away by the escaping fragment.

This 'thought experiment' opened a new set of perspectives to science fiction writers. Figure 54 was inspired by a book by Charles Misner, Kip Thorne and John Wheeler[4] which is devoted to gravitation and is a kind of 'bible' of General Relativity. The idea consists of using the ergosphere of a rotating black hole to solve the energy problems of an advanced civilisation. It involves constructing a vast rigid structure around the black hole, at a sufficiently large distance to avoid tidal effects. An industrial city is built on the structure. Every day millions of tonnes of rubbish are collected in skips and tipped down a hole in the structure. From there the skips are dispatched in the direction of the black hole, one after another. Each skip is sent in a descending spiral. When it penetrates the ergosphere and reaches an 'ejection point', an automatic mechanism opens the skip and expels the rubbish in a carefully calculated retrograde orbit. The rubbish captured by the black hole slightly decreases its rotational velocity. At the same time the empty skip leaves from the ergosphere with an increased energy. It is finally recovered by a giant rotor, where it deposits its colossal kinetic energy. The rotor is linked to an electric generator which supplies the city with electricity. For each recovered skip the net gain in energy is equal to the mass-energy of the ejected rubbish, plus a fraction of the mass-energy of the black hole itself. So by this clever mechanism the inhabitants of the city have not only converted all the mass of their rubbish into electric energy but also a fraction of the total mass-energy of the black hole. An ecological triumph!

[4] Misner, C. W., Thorne, K. W., Wheeler, J. A.: *Gravitation*, 1973, W. H. Freeman and Co. (San Fransisco).

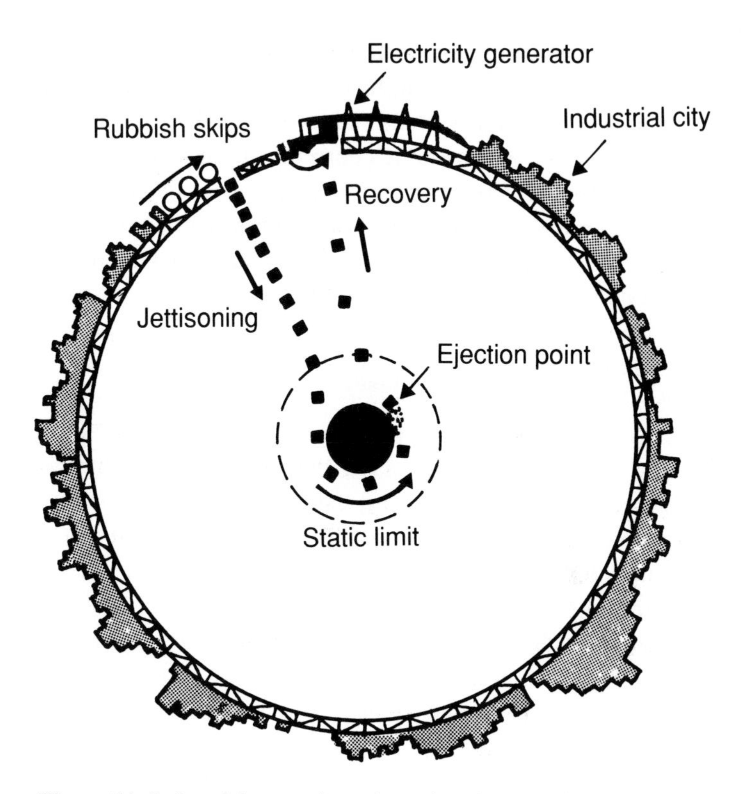

Figure 54. Industrial extraction of rotational energy from a black hole.

The black hole dynamo

Penrose's mechanism may be more than just an anecdote. It may be that the extraction of rotational energy from a black hole already occurs in natural astrophysical conditions, via a suitably arranged external magnetic field.

The French astrophysicist Thibaut Damour has drawn an analogy between the surface of a black hole and a moving charged soap bubble. Specifically, the black hole is a conductor of electricity characterised by a certain electrical resistance. Thus when a rotating black hole is placed in an electric field it acts like an electric motor, because of the *dynamo* effect. As in a gigantic electromagnet, the phenomenon of induction between the rotor (black hole) and the stator (external magnetic field) creates circulating electric currents on the horizon capable of slowing down the rotation and extracting some of its energy. These induced currents are analogous to 'Foucault's currents', used in the braking systems of some heavy vehicles.

Favourable conditions for the extraction of energy from a black hole by the dynamo effect may perhaps exist in the centres of certain galaxies containing giant black holes, (see Chapter 17).

The black hole laser

Another way of extracting rotational energy from a black hole was suggested in 1971 by the Russian physicist Yacov Zeldovich. The mechanism, called *superradiance*, is based on an analogy with a well known quantum mechanical phenomenon: stimulated emission of particles.

In an atom, the electrons occupy orbits whose energy is quantised, that is, equal to multiples of a fundamental unit. Lower orbits have lower energies. In a 'normal' atom the electrons tend to occupy the lower orbits. This is why an electron occupying a higher level can spontaneously jump to a lower level by emitting a photon – a particle associated with an electromagnetic wave – whose frequency corresponds to the difference between the upper and lower energy levels. This is called *spontaneous emission.*

Conversely, if an atom is 'illuminated' by an electromagnetic wave of suitable frequency, the wave causes electron transitions from low energy levels to higher ones. The wave is partially absorbed by the atom and is retransmitted with a lower energy. Now let us imagine that in a suitably prepared atom most of the electrons already occupy high levels; we say that the atom is in an *excited state*. In this case the in-coming electromagnetic wave can only produce transitions from high levels to lower ones. This is called *stimulated emission*, and the wave is now *amplified* during the interaction, gaining energy. This mechanism, discovered by Einstein in 1916, is how the *laser* funtions. The laser is one of the most beautiful technological advances based on the quantum properties of matter and radiation.

A similar mechanism could be produced using a rotating or

charged black hole (a Kerr–Newman black hole). This can be considered as an 'excited state' of the static and neutral Schwarzschild black hole. We have already seen in Chapter 10 how an illuminated black hole is capable of absorbing and partially reflecting incident light. However, when the discontinuous properties of radiation are taken into account, new effects appear and reveal the links between gravitation and quantum physics. If electromagnetic or gravitational waves of a suitable frequency and phase are directed towards a Kerr–Newman black hole, the reflected waves are amplified. In other words, the black hole gives up energy to the scattered waves. This phenomenon of *superradiance* would in principle enable us to extract rotational or electrical energy from a black hole.

Let us follow further the analogy between a Kerr–Newman black hole and an excited atom. Since a black hole allows stimulated emission, it should also allow the spontaneous emission of particles. As it is (classically) forbidden for a particle to leave the event horizon, the spontaneous creation of particles must occur outside the black hole.

This intuition is verified by detailed calculations of the interaction of a black hole, described by General Relativity, with matter or radiation, described by quantum mechanics. The 'de-excitation' of a black hole can appears as a neutralisation of its charge through the emission of particles of the same sign, and a slowing down of its rotation by the emission of particles whose spin is in the same sense as the angular momentum of the black hole. In principle, all types of particles could be created (photons, neutrons, electrons, protons and so on), but the larger the particle the less chance it has of being produced.

In this fashion the developments of thermodynamics of black holes have taken us to the frontier between the 'classical' world and the 'quantum' world. On the way we have discovered that black holes have more properties than the passive gravitational wells they might appear to be at first sight. The arrival of the *quantum black hole* in 1974 was to confirm its black colour but remove its last classical property: that of being a hole.

14

The quantum black hole

'There is always a moment when curiosity becomes a sin; the devil always stands next to the scientists.'

Anatole France

The shrinking black hole

In 1971, Stephen Hawking suggested the existence of *mini black holes*. According to Hawking, in the first moments of the Universe, well before the birth of stars and galaxies, the pressure and energy of the ambient 'cosmic bath' were so great that they would have been able to force small lumps of matter to concentrate into black holes of various sizes and masses.[1] In particular, minuscule black holes could have formed, having the mass of a mountain and the size of an elementary particle. These would differ from black holes forming in the present Universe, which require the gravitational collapse of large amounts of matter.

Hawking then considered the interactions between these mini black holes and the ambient medium. In this case the distances involved are microscopic, and matter and energy should be described by quantum mechanics. As already indicated, there is no satisfactory theory of quantum gravity. However, the gravitational field, including space-time itself, does not really become discontinuous until we reach the Planck length, which is very much smaller than the radius of an elementary particle or a mini black hole. The interaction between a mini black hole and the ambient matter and energy can therefore be calculated on the basis of a compromise: the space-time continuum remains 'classical' and can

[1] See Chapter 15.

be described by General Relativity, and only its content – matter and energy – is quantised.

Hawking used this approach in his calculations in 1974, and discovered a phenomenon so unexpected that he thought he had made a mistake, and checked his calculations several times. Eventually he was forced to accept the result: *a mini black hole must evaporate by emitting particles*!

This is very disconcerting at first glance: such behaviour is a flagrant contradiction to the 'classical' conception of a black hole that forbids the escape of anything from the event horizon. Of course an 'excited' black hole may lose some of its energy by slowly decreasing its angular momentum or charge, but in that case the particles are emitted outside the horizon. A 'de-excited' Schwarzschild black hole is forced to maintain an irreducible mass-energy, related to its area and entropy, which from the second law of classical thermodynamics can only increase over time. Now Hawking's calculations showed that a mini black hole, excited or otherwise, has to allow particles to escape and evaporate by losing its mass and energy. How is this conflict to be resolved?

It is often easy to interpret a great theoretical discovery with the benefit of hindsight simply because it suddenly explains the relationship between little understood phenomena. In this sense, the quantum evaporation of black holes arrived just at the right moment to justify completely the thermodynamic picture of black holes. For, when closely examined this, in its 'classical' version, is *inconsistent*. Let us see why.

According to the laws of thermodynamics, all bodies at a certain temperature immersed in a colder medium, such as air, must lose energy by radiation. As this happens their entropy decreases while the entropy of the external medium increases. During this exchange, the *total* entropy – the sum of the individual entropies – must increase, by virtue of the second law.

What does thermodynamics say about a black hole? It has an entropy, given by its area, and a temperature, given by its surface gravity. Let us put the black hole in a heat bath. If the hole is cooler than the heat bath, it will absorb energy and increase its entropy. However, if the black hole is hotter than the bath we would have to believe that it should lose energy and entropy to it, contradicting the second law of 'classical' black hole thermodynamics.

The inconsistency was removed by Hawking's discovery. Thanks to certain properties of quantum mechanics, which I shall explain later on, a black hole is able to emit particles or radiation even if it is in a state of minimum energy. By losing energy, the black hole decreases its entropy – that is, its area – while the entropy of the heat bath gaining the energy increases. In fact the increase of entropy in the bath is greater than the loss of entropy in the black hole. Consequently the second law of thermodynamics is satisfied by the full system of black hole + bath, whose entropy always increases.

The tunnel

Classically nothing can leave a black hole; the event horizon acts as a sort of 'one-way membrane', allowing matter in and preventing anything from escaping. From the inside, the event horizon appears as an infinitely high wall, which to jump over requires an infinite amount of energy.

But quantum mechanics offers the possibility of crossing any wall at all, even without enough energy. This phenomenon, called the *tunnel effect*, is a direct consequence of the Uncertainty Principle, which is the touchstone of quantum mechanics, much as the Principle of Equivalence is for General Relativity.

In quantum mechanics we learn that there is some 'fuzziness' in the microscopic description of the world. If, for example, we want to measure the position of an isolated electron it must be localised and visible. In order to be visible, it must be illuminated. Now an electron is so small that the photons which are used to illuminate it also cause it to move: the photons give the electron a small impulse and modify its velocity. Measuring the position of an electron to high precision therefore introduces a degree of *uncertainty* into the measurement of its velocity. If the velocity of an electron is known to an accuracy of 1 cm/s, it is impossible to localise it more exactly than within 1 centimetre.

More generally, all measurements perturb the microscopic system. The Uncertainty Principle was formulated by Werner

Heisenberg in 1927. Of course quantum indeterminacy is diminished when the masses involved are much greater. Thus a proton, which is about 2000 times more massive than an electron, can be localised to within about 5 micrometres if its velocity is known to a precision of 1 cm/s. This precision, although better, is still poor when we remember that the diameter of the proton is a billion times smaller. However, the masses of macroscopic objects are so enormous compared with those of elementary particles that the uncertainty over their position and momentum disappears completely. The macroscopic world is 'deterministic'.[2]

The Uncertainty Principle can be applied to other physical parameters which are quantised, such as energy: in a very brief time interval energy can fluctuate by a certain amount. Classically, escape from a black hole is forbidden, but the Uncertainty Principle allows the particle to borrow a certain amount of energy from the black hole for a certain time interval. If the black hole is microscopic, about the size of an elementary particle, the 'jump' of energy may be sufficient to move a particle a distance greater than the radius of the horizon. The result of this operation is a loss of energy to the black hole through the escape of a particle. The particle has not really jumped over the event horizon; rather it has passed through a 'tunnel' briefly opened by the Uncertainty Principle.

Vacuum polarisation

There is an equivalent interpretation of black hole evaporation, in terms of what is called *vacuum polarisation*.

In quantum mechanics, the vacuum does not mean the absence of any field, particle or energy. The quantum vacuum is the state of minimum energy; it is only called 'vacuum' because a state where the energy is precisely zero cannot exist.

[2] Contrary to currently held belief, this does not mean that its evolution can be predicted. Many very complicated but perfectly classical ('nonlinear') physical phenomena, although governed by deterministic equations, evolve towards totally unpredictable states. This is the reason why weather forecasts are so unreliable when predicting more than a week in advance, no matter how powerful the computers used!

The Uncertainty Principle for time and energy explains why the quantum vacuum is populated. By virtue of mass-energy equivalence, an energy fluctuation in the vacuum can cause elementary particles to appear. In 1928 Paul Dirac discovered that each elementary particle had a corresponding *antiparticle* with the same mass but 'mirror' properties. Thus the electron has a negative electric charge, and its antiparticle, the positron, has the same mass but opposite electric charge. The photon, which has no mass, is its own antiparticle. If a particle and its antiparticle meet they annihilate each other, transforming their mass into energy. A particle and its antiparticle therefore represent a quantity of energy equal to twice the rest mass, and conversely a quantity of energy can be regarded as a set of particle–antiparticle pairs. Thus the quantum vacuum, agitated by energy fluctuations, can be regarded as a 'Dirac sea', populated by particle–antiparticle pairs spontaneously appearing out of the vacuum and annihilating themselves shortly afterwards.

In 10^{-21} seconds, an electron–positron pair can spontaneously appear and disappear. Pairs of heavier particles can also appear in the vacuum, but by the Uncertainty Principle they can exist only for much shorter times. A proton–antiproton pair created in the vacuum survives on average for a time 2000 times shorter than an electron–positron pair.

In the quantum vacuum in the absence of all forces, pairs are continuously being created and destroyed, so that on average no particle or antiparticle is truly created or destroyed, nor can they be directly observed. The pairs are said to be *virtual.* Imagine now a force field, such as an electric field, imposed on the vacuum. When an electron–positron pair appears from the vacuum, the electron and positron are deviated in opposite directions by the field. If the electric field is strong enough the pair separate so much that they cannot collide and annihilate each other. The particles now become *real,* and we say that the vacuum is *polarised.*

The spontaneous creation of particles by polarisation of the vacuum is not a theoretical fantasy, but a phenomenon verified in the laboratory. Let us consider a hydrogen atom in the quantum vacuum. It consists of a negatively charged electron and a positively charged proton. All around, virtual pairs of particles appear and disappear continuously, but the electric field created by the proton and electron polarises the vacuum in their immediate neighbourhood. The oppositely charged particles tend to separate, and for a brief instant tiny electric currents are produced. These make the electron wobble in its orbit. This causes a slight shift in the radiation emitted by the hydrogen atom, called the 'Lamb shift', which was experimentally detected in 1947.

However, the vacuum is not easy to polarise. A high energy density is needed to cause the virtual pairs to separate and real particles to appear. The nature of the energy involved is not important. It could be electric: when the voltage between the plates of a condenser exceeds a certain limit, the vacuum polarises and the condenser 'clicks'. It could be thermal energy: a piece of metal heated slightly emits photons (which are their own antiparticles), but at a trillion K it radiates electron–positron pairs.

Since all forms of energy are equivalent to mass, it is logical to expect that *gravitational energy* can also be spontaneously converted into particles. This is precisely the deep meaning of Hawking's discovery. The quantum vacuum is polarised by the very intense gravitational field surrounding the mini black hole (Figure 55). In the Dirac sea, virtual pairs are constantly being created and destroyed. For a brief instant a particle and its antiparticle separate. There are four possibilities: the two partners come together again and annihilate each other (process I); the antiparticle is captured by the black hole and the particle materialises in the external world (process II); the particle is captured and the antiparticle escapes (process III); both partners plunge into the black hole (process IV). Hawking calculated the probabilities of these processes occurring and discovered that process II was the most common. The energy balance is thus as follows: by preferentially capturing the antiparticles, the black hole spontaneously loses energy and therefore mass. To the external observer, the black hole appears to evaporate by emitting a stream of particles.

Black is black

We have now examined all the mechanisms for extracting energy from a black hole. The rotational and electrical energy can be removed by both classical and quantum processes. In particular, the de-excitation of a charged and rotating mini black hole by superradiance discussed earlier can also be reinterpreted in terms of vacuum polarisation. From the virtual pairs surrounding them, black holes prefer to capture particles with the opposite sign of charge or angular momentum to their own. Thus even in the vacuum a mini black hole initially formed with a non-zero charge and angular momentum has a tendency to neutralise itself spontaneously and slow down, rapidly trying to attain the Schwarzschild state. However, the Schwarzschild state loses its classical 'irreducibility' and the 'inert' mass spontaneously evaporates. What is the exact nature of the emitted radiation?

By a curious irony, the black hole radiates like the other physical paradigm of the same 'colour', the *black body*. A black body is a sort of ideal radiator, in perfect thermal equilibrium characterised by a certain temperature. It emits at all wavelengths, with a spectrum which depends only on its temperature and not on its detailed nature. A perfectly opaque furnace heated to a given temperature, with a small hole pierced in the side for the observer to view the radiation, gives an approximate idea of a black body. The black body was in fact one of the historical sources of quantum mechanics. In 1899 Max Planck produced his hypothesis of energy quanta whilst studying its properties.

Hawking's calculations demonstrate that the evaporation radiation of a black hole has all the characteristics of a black body. This result renders the thermodynamics of the black hole completely consistent, by assigning it a true temperature, uniform over the horizon, given directly by the surface gravity.

For the Schwarzschild black hole, the temperature is inversely proportional to the mass. If the black hole has the same mass as the Sun its temperature is negligible: a ten millionth of a degree Kelvin (above absolute zero). This is not surprising since the evaporation phenomenon is of quantum origin and concerns mini black holes in particular. These are actually very hot. A black hole having the mass of a small asteroid has the temperature of a 'white hot' furnace (6000 K) and radiates in the visible. A 'typical' mini black hole of 10^{15} grams, the size of a proton, has a temperature of a trillion K. At this energy the radiation is not in the visible range, but consists

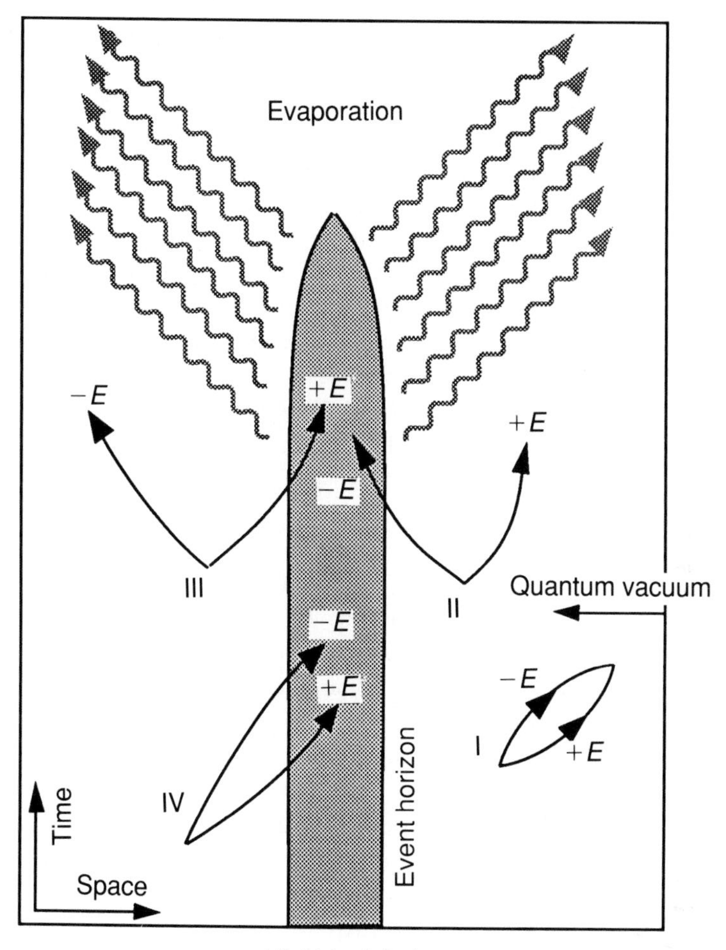

Figure 55. The quantum evaporation of a mini black hole by polarisation of the vacuum.

of a mixture of gamma ray photons and massive elementary particles.

The smaller the black hole the hotter it is. The emission from a mini black hole can therefore only increase, and the last stages of evaporation appear like a cataclysmic explosion. The 10^{15} gram black hole would take 10 billion years to evaporate completely, but during the last tenth of a second it would liberate as much energy as a million 1 megatonne hydrogen bombs.

The final result of the evaporation of a black hole is unknown. It might be argued that the disappearance of the event horizon would leave a naked central singularity, but this classical vision is probably wrong. When the radius of the black hole is reduced to the order of a Planck length (10^{-33} centimetres), the quantum fluctuations of the space-time geometry itself become important and only a quantum theory of gravity could decide the ultimate fate of a mini black hole. If it evaporated completely by radiating all its mass, flat space-time should logically result. Quantum gravity is the royal road for understanding both the Big Bang and the destiny of black holes, the beginning and end of the Universe.

Gravitational instability

A conventional thermodynamic system immersed in a colder medium loses energy. Its temperature decreases and the medium's increases until an equilibrium is reached. We say that it has a *positive specific heat*. The quantum black hole behaves as the inverse. Its temperature increases as it loses energy and vice versa. Placed in a hot medium, the black hole has a tendency to absorb energy and increase in size, thus cooling, until all the available energy has been absorbed. If on the other hand the black hole is placed in a cooler medium, it radiates, reducing in size, ending by evaporating and dispersing all its energy. The black hole has a *negative specific heat*. It is therefore fundamentally unstable.

All self-gravitating systems – that is those whose equilibrium depends only on gravitation – are unstable whether they are quantum systems or not. For example, an artificial satellite orbiting the Earth slowly loses its gravitational energy to atmospheric friction and starts to spiral towards the Earth. As this happens, its velocity and therefore its kinetic energy increase, so that it finally crashes into the Earth without ever finding a stable orbit.

Gravitational collapse is an extreme example. As it collapses under its own weight, an ensemble of particles such as a star or stellar cluster radiates its gravitational binding energy and contracts, becoming hotter and hotter. In the absence of any opposing force, the inevitable formation of a singularity shows the impossibility of attaining an equilibrium state. The evaporation of a mini black hole is no more than gravitational collapse in the opposite direction, as can be verified by returning to the space-time diagram of Figure 55. As matter leaves the horizon the 'instantaneous' state of an evaporating mini black hole is that of a white hole! Thus quantum mechanics provides black holes with the property of instability characteristic of gravitation in general.

Furthermore, the link between gravitation and thermodynamics is probably a characteristic of nature, embracing domains much wider than black holes. Actually, in the thermodynamic transformations of a black hole the crucial role is played by the event horizon. Now event horizons also exist in contexts far removed from black holes. In the flat space-time of Special Relativity, with no gravitation, an observer with constant acceleration cannot 'classically' have access to information coming from a distant region of space-time simply because the radiation emitted there can never be 'caught'. For him all this part of space-time is hidden behind an event horizon. If we take into account the quantum fluctuations in the vacuum it can be deduced that the *acceleration*[3] *polarises the vacuum*. If the observer carries with him a particle detector he will measure a 'quantum noise' in the form of radiation from a black body whose temperature is proportional to his acceleration. In cosmology, models of the expanding Universe possess event horizons. These also have an associated black body temperature.[4]

The thermodynamics of black holes has taken us a long way from steam engines.

God cheats

Elementary particles interact through nuclear and electromagnetic forces. These interactions obey certain rules which have been verified by observation and have allowed scientists to build a

3 Otherwise equivalent to a uniform gravitational field, see Chapter 3.

4 Extremely low, and not to be confused with the temperature of the cosmic background at 2.7 K, the remnant of the Big Bang.

coherent theory of particle physics. One of these rules is the *conservation of baryon number.* In simple terms it states that in all fundamental interactions the proportion of particles and antiparticles must be conserved. Thus a photon (baryon number 0) can be transformed into a pair consisting of a neutron (baryon number +1) and an antineutron (baryon number −1), because the total baryon number remains 0. On the other hand, a neutron can never be converted into a photon pair. A similar rule applies to another family of particles called leptons, which include electrons, muons and neutrinos. Each of these has a lepton number which must be conserved during fundamental interactions.

This basic principle of particle physics is cheerfully violated by the quantum black hole. We have already seen that when a black hole is formed or swallows matter it 'loses its hair': all information about the particles is lost when they pass through the event horizon. In particular, a black hole formed from baryons (for example the protons and neutrons at the heart of a massive star) does not remember its baryon number. It could just as well have been formed from antibaryons, without our being able to tell the difference. Let us wait patiently. After a time the black hole starts to radiate according to Hawking's mechanism, releasing energy and entropy. Now the fact that the black hole radiates like a black body means that it can emit only *equal* numbers of baryons and antibaryons, or leptons and antileptons. In other words, the net baryon number leaving the evaporating black hole is always zero. The evaporation of the black hole violates the rule of the conservation of baryon and lepton numbers.

This astonishing property is a good illustration of the manner in which information released to the external medium by the evaporation of a black hole 'degrades' as it passes through the horizon. This deterioration puts a 'thermal stamp' on the matter and radiation leaving the hole and randomises the data. For this reason, Hawking considered that the Uncertainty Principle applied to the black hole was replaced by what he called the '*Unpredictability Principle*'.

Einstein never liked quantum mechanics, although he played a pioneer role in its development. He disliked the idea of indeterminacy inherent in the Uncertainty Principle, and he expressed his aversion in the phrase, 'God does not play dice.' Hawking's response was, 'Not only does He play dice, but He throws them where we cannot see them!'

PART 4
LIGHT REGAINED

'The true mystery of the world is not the invisible but the visible.'

Oscar Wilde

15

Primordial black holes

Lumps

Let us consider the very distant past of the Universe, 15 billion years ago. The Universe was still newly formed and not a smooth and homogeneous soup with a well-defined structure, but a 'quivering' mass agitated by slight fluctuations which, under the influence of their own gravitation, tended to cluster together to form 'lumps'. However, like a cake which rises in the oven, the Universe was expanding under the force of the Big Bang. The opposition of the general expansion to the localised condensations raises one of the biggest questions of contemporary physics: how did some of these lumps go on to form galaxies? After all, the expansion of the Universe should have inhibited local condensations so that no galaxy, star or planet – and at the far end of the chain no living creature – should ever have appeared in the history of the Universe.

The existence of galaxies proves *a posteriori* that some of the fluctuations in the primordial universe were able to grow and dissociate themselves from the universal expansion. During such condensations the density contrast, that is, the excess of mass in the lump compared to the ambient density, would increase without limit. At the beginning it was tiny, scarcely one part in a thousand, although already the equivalent of several hundred $M_\odot$. Nowadays, for the same mass, the density contrast would have to be more than one to a hundred thousand. Gravitation has worked hard.[1]

[1] The contrast in density between a solar type star and the interstellar medium is even greater: one to 10^{30}.

All cooks know that when stirring a sauce over heat, it is easier to form small lumps than big ones. It is therefore possible that high-amplitude fluctuations in the primordial Universe condensing masses much smaller than galaxies could have created gravitationally condensed bodies in the first place. It was by invoking a mechanism like this that in 1971 Stephen Hawking suggested the existence of *primordial black holes*.

The reader will recall that the mass of a black hole formed by the collapse of a star is of the order of 3 $M_{\odot}$. There is no such constraint on primordial black holes. Black holes of all shapes and sizes could have condensed out in the early period of the Universe, in particular *mini black holes* the size of an elementary particle.

Is it possible to test the idea of mini black holes by astronomical observation?

Worlds in collision

Far and away the best way of detecting mini black holes would be finding one in the Solar System. Hawking suggested that a mini black hole could be captured by the Sun and fall gradually towards its centre. Contrary to popular belief the Sun would not be consumed by it. A small black hole would in fact be able to exist for a long time without having any noticeable effect on the Sun. The Sun would only be in any danger if the black hole grew very quickly. However, the solar matter swallowed by the black hole would emit so much radiation before disappearing that the radiation pressure exerted on the external medium would limit the speed of growth of the black hole. The flow of consumed matter would adjust itself to the flow of liberated energy, so that the region around the black hole would become nothing less than an extremely stable nuclear reactor. The Sun with the 'black heart' would continue its peaceful life on the Main Sequence, its functioning modified in a barely perceptible way!

This rather unusual scenario has been invoked to explain the disagreements between the measured quantity of solar neutrinos reaching the Earth and the number predicted by the theory of nuclear reactions. It has since been abandoned in favour of more conventional mechanisms which are better at explaining the differences.[2]

It goes without saying that a collision between a mini black hole and our planet is extremely unlikely; there is more chance of us colliding with a big meteorite. Such an event has nevertheless been used as one of the possible explanations for the famous catastrophe at Tunguska in Russia. On 30 June 1908 the Yenissey Valley in Siberia was devastated by the fall of a celestial body. The explosion was accompanied by optical, acoustic and mechanical phenomena; the shock wave ravaged the forest for several kilometres, killing hundreds of reindeer and being audible for more than 1000 kilometres, breaking windows and shaking buildings. As registered by seismographs, it was equivalent to 1500 of the bombs dropped on Hiroshima. The sky was lit up and for a period its brightness was sufficient for reading a book in the middle of the night in the Caucasus. However, the site of the explosion was only scientifically examined 20 years later. The trees had been burnt over a radius of 15 kilometres and flattened over a radius of 30 kilometres, all lying facing away from the centre of the explosion, although no crater marked the point of impact.

Many reasons ranging from the banal to the bizarre were suggested as the cause of the catastrophe. The explanation adopted today suggests a meteor, or more precisely, a cometary fragment. A piece of ice and rock measuring hundreds of metres and falling in the opposite direction to the Earth's rotation at a speed of 50 km/s would reproduce the effects seen at Tunguska: evaporation in the atmosphere with the injection of numerous particles would not leave a crater or sizeable residues. The best proof comes from the chemical analysis of small pieces of debris recovered from the site: they consist mostly of silicates and nuggets of ferronickel, whose composition conforms perfectly to that of comets.

This evidence did not prevent two American astrophysicists from proposing a radically different explanation, that a mini black hole had passed through the Earth like a hot knife through butter and

[2] It could happen, for example, that neutrinos have a non-zero mass, so that the theoretical flux, usually calculated assuming zero mass, becomes compatible with the observed value.

come out on the opposite side of the Earth from Tunguska, which happens to be in middle of the South Atlantic Ocean, where no tree or window could bear witness to what happened.

A deeper analysis shows that the passage of a black hole through the planet would cause seismic waves which were not observed, and that its exit would be accompanied by atmospheric shock waves which were also not observed. The explanation of an errant mini black hole was fairly fantastic[3] but it was a good piece of publicity. However, specialists in black holes have found little good in it; seeing black holes everywhere does little to reinforce their credibility.

A brief life

The main hope of detecting mini black holes lies in the basic property, discovered by Hawking, on the basis of quantum mechanics: the *evaporation of their mass* as black body radiation.

The theory of density fluctuations shows that low mass black holes could only have been formed during the first moments of the Universe. However, mini black holes are condemned to evaporate more rapidly the smaller they are.[4] A black hole weighing one tonne would evaporate in a ten billionth of a second, while a black hole weighing a million tonnes would survive for 10 years. The only primordial black holes which could have survived until today are those with lives longer than the current age of the Universe, 15 billion years. The minimum mass for these would be a *billion tonnes*. This is roughly the mass of a mountain, and the corresponding black holes would have a radius of 10^{-13} centimetres, the same as a proton!

As to the more massive black holes, they can evaporate only in a much longer time than the age of the Universe. The lifetime of a 1 $M_\odot$ black hole is about 10^{66} years, for example. This enormous number should not be surprising, because evaporation is a quantum phenomenon and so occurs over very small distances,

3 Less, however, than those invoking a piece of antimatter or the disintegration of a flying saucer in distress!

4 The lifetime of a black hole is proportional to the cube of its mass.

comparable to those of elementary particle diameters. Thus evaporation is completely insignificant for black holes which are more massive than a mountain, whether they were formed at the beginning of the Universe or later during the explosions of supernovae. In fact for all big black holes the rate at which they increase in size exceeds the rate at which they evaporate. We should now ask what mass will currently evaporating black holes have.

In answering this we should remember that black holes do not exist in a perfect vacuum, but in a material medium which possesses a certain energy, at least equal to that of the cosmic microwave background radiation, the remains of the primordial explosion. The temperature of this cosmic 'bath' is 2.7 K. The laws of thermodynamics imply that, of the surviving primordial black holes, only those smaller than 10^{26} grams (the mass of the Moon, with a radius of 0.1 millimetre) have a temperature greater than 2.7 K, and consequently should evaporate by giving energy to the ambient medium. Larger black holes should on the other hand absorb cosmic energy and increase in size. We conclude that black holes smaller than 10^{15} grams have already evaporated, those between 10^{15} and 10^{26} grams are in the process of evaporating, and those greater than 10^{26} grams, including the 'second generation' of stellar black holes, are in the process of increasing in size.

Le dernier cri

How can we observe an evaporating black hole of suitable mass? Hawking's calculations show that during the last tenth of a second of its existence the evaporation becomes explosive, suddenly destroying the black hole by converting its mass into energy. This energy is dissipated in an intense burst of gamma rays, which in principle at least should be detectable within a radius of 30 light years.

Referring again to Table 1,[5] we note that gamma radiation on average transports a million times more energy than visible radiation. It is therefore highly penetrating and would be fatal to terrestrial life if not blocked by the upper layers of the Earth's

5 See page 13.

atmosphere. One way of observing cosmic gamma radiation is to use the atmosphere itself as the detector. As they pass through the upper atmospheric layers, the gamma-ray photons convert their energy into matter, creating showers of particles and antiparticles. At the instant that they are created, these particles propagate at exactly the velocity of light *in a vacuum*; therefore they are travelling faster than light through the *air*. This sudden injection of 'super-relativistic' particles into the Earth's electromagnetic field is similar to an aeroplane breaking the sound barrier: they cause a shock wave, which instead of producing a sonic bang causes a flash of visible light called *Cerenkov radiation*. This type of radiation is easily detectable on the ground, and has long been used to measure the flux of gamma radiation reaching the Earth from the cosmos.

These bursts of gamma radiation which reach the Earth at a rate of several per year and are detected by their Cerenkov 'light' do not, however, possess the characteristics of the explosions of mini black holes. It is clear that mini black holes are not the only celestial sources of gamma radiation. Besides, apart from these sudden bursts of radiation, there is a lower level of continuous gamma radiation which has been measured by detectors on satellites orbiting above the atmosphere. This important discovery shows that many astronomical phenomena inject high energy radiation into interstellar space. The detailed origins of this diffuse background gamma radiation is somewhat controversial, but the general belief is that it is caused by compact stars such as neutron stars (see Chapter 16) or on a much greater scale by active galactic nuclei.

Nevertheless, many mini black holes could have exploded in the recent past and contributed part of this radiation. The SAS2 satellite carried out precise measurements of the diffuse gamma radiation flux; its level is so low that even supposing *all* the observed flux came from such explosions, the average number of primordial black holes contained in a volume of one cubic light year could not be more than 200. In these conditions the mini black hole closest to the Earth would be situated very far from the Solar System.

The true density of primordial black holes is much smaller. Stricter constraints than that imposed by gamma rays have been proposed. During an explosion of a mini black hole the emitted particles would interact with the general magnetic field in the

Galaxy and produce characteristic radio waves. Since radio waves are much easier to detect than gamma radiation, the explosions of mini black holes should be detectable by giant radio telescopes. The fact that none have ever been detected imposes a very severe limitation on the frequency of such explosions: it must not exceed one per 3 million years in a volume of 1 cubic light year.

Primordial mini black holes with the mass of a mountain may exist, but they are extremely rare.

Gravitational mirages

The absence of characteristic traces of the explosions of mini black holes does not preclude the existence of primordial black holes more massive than 10^{15} grams, which have still not evaporated. Can they be detected?

We need only recall the 'illumination' experiments described in Chapter 10 to convince ourselves that even a completely isolated black hole can focus radiation from distant sources by acting as a 'gravitational lens'.

Suppose that the Earth, a black hole and a distant star are fortuitously aligned. According to the laws of General Relativity the curvature of space-time near a black hole causes light from a distant star to follow one of several possible trajectories before reaching the Earth (Figure 56). Under these conditions telescopes should be able to see several images of the same source: a 'main' image corresponding to the least-deviated light rays, and ghost images corresponding to the more 'twisted' light. This displacement of the apparent images with respect to the real image is called a *gravitational mirage*.

The traditional mirage sometimes observed in the desert is caused by heat from the sand which by conduction heats up different layers of air to different temperatures, giving them different refractive indices. The light rays reflected off the sand follow several different possible trajectories before reaching the distant traveller. We can imagine how the mysterious phantom images might be interpreted as an oasis, a town or an expanse of water, depending on what the traveller would most like to see.

Undoubtedly the gravitational mirages caused by the distortion of cosmic space are much more difficult to detect. Let us consider the example of a giant black hole outside the Galaxy. Distant sources such as *quasars* or cosmic background radiation[6] might be affected by the black hole's gravitational lensing effect.

Astronomers are already aware of a score of cases where gravitational mirages give multiple images of quasars. However, these mirages are caused not by giant black holes but more prosaically by intervening galaxies. We recall that all concentrations of matter deform space-time continuum to some extent and can act as gravitational lenses. Most measurements (image separation, etc.) only give us the mass of the lens, so if the lens remains undetectable it is impossible to say if we are dealing with a giant black hole or a faint galaxy.

In 1985 a pair of quasars called Hazard 1146 + 111 B and C caused a sensation among the astronomical community. Their redshifts were at first sight identical, and it was tempting to interpret them as images of the same object, doubled by an intermediary lens. But Hazard 1146 + 111 differs from other gravitational mirages in the extremely large angular separation between the components: 2.6 arcminutes, which is 20 times greater than the other multiple quasars known to date. If these were images of the same object the gravitational lens would require a mass equivalent to several thousand galaxies.

Three types of objects could act as a massive lens of this type: an extremely dense cluster of galaxies, a 'supergiant' black hole, or a 'cosmic string'. No observational measurements revealed the existence of a galaxy cluster situated in that direction. 'Cosmic strings' are elegant structures invented by elementary particle theorists; they would have been produced during the first instants of the Universe and consist of long strings with almost zero radius which transport gravitational energy. However, there is no experimental procedure which can be used to confirm their existence or the validity of theories underlying them. Finally, there was the hypothesis of the black hole, which paradoxically was the least 'exotic' explanation. The black hole in question would have

[6] The cosmic background radiation is indeed the only source of electromagnetic radiation present everywhere in the sky.

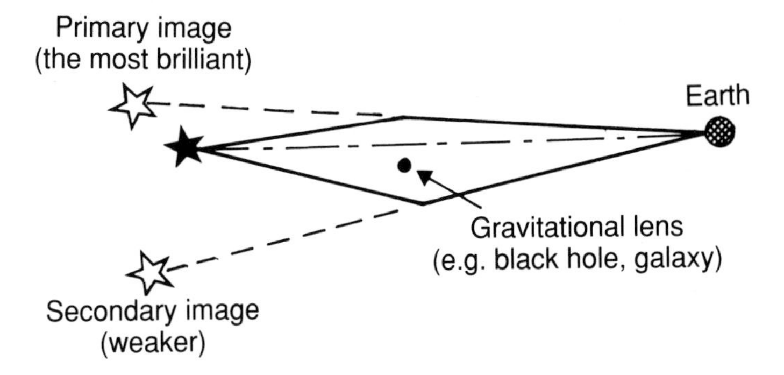

Figure 56. A gravitational mirage.

had a mass of between 10^{12} and 10^{17} $M_{\odot}$, and could only have been of primordial origin; its colossal mass would have far exceeded anything imagined by black hole scientists.

But before accepting this extreme view, one had to be sure that Hazard 1146 + 111 is in fact a gravitational mirage. More precise measurements were made, and showed that the spectra were different; in other words, the two images did not come from a unique single quasar, but two physically distinct quasars very close to each other. This was the end for dreams of strings and super black holes. I have recounted this anecdote in detail to emphasise the fact that scientific research teems with this kind of confusion. When a sensational discovery is announced (and attracts the attention of the media), it often results from erroneous interpretations of imprecise data. Better measurements then put the event back into the category of the 'normal', and show again the relevance of the Principle of Simplicity : the most 'economical' hypothesis, i.e. the most 'conventional' (without pejorative overtones), is almost always correct.

After giant black holes, let us now consider the case of a stellar-mass black hole (primordial or not), isolated in our Galaxy, and thus fairly close. As the hole has a diameter of a few kilometres only, its apparent diameter at a distance of several tens of light years is so minuscule that the probability of an alignment with a more distant star in our Galaxy is incredibly small. In fact, even if this alignment occurred, the angular separation between the various stellar images, which depends on the mass of the black hole, would be well below the resolving power of present and future telescopes. Is the situation then hopeless? No, since the effect of a lens, even a microscopic one, is not confined to multiplying the image, but also amplifies the intensity of the image and distorts its spectrum. Let us thus consider a microlens in the halo of our Galaxy or that of a nearby galaxy, which is therefore moving very slightly in projection over the distant (therefore fixed) background of the quasars. The probability of an alignment is no longer negligible. The mirage thus produces a brief change in the luminosity and the spectrum of the quasar. The idea is so well founded that an entire class of galaxies with active nuclei (see page 274) is interpreted by some workers as resulting from the cumulative effects of microlensing. Several intensive observational programmes are under way. The aim is not to detect stellar black holes so much as to reveal the existence of a large number of very small faint stars populating the haloes of galaxies.

Dark matter

One of the unresolved problems of modern cosmology is that of *missing mass*. Observations of the motion of galaxies indicate that the 'visible' matter (visible in the optical, radio, infrared and X-ray domains) forms only a fraction of the total mass. A simple example will illustrate the problem. Many observed galaxies group together in clusters, forming bound gravitational structures which do not merge into the surrounding cosmic medium. Now if these clusters consisted only of the detectable individual galaxies and intergalactic gas, their gravitational attractions would not be strong enough to keep them together. There must therefore exist some *dark matter*, invisible in electromagnetic radiation but which acts gravitationally to hold the galaxy clusters together.

Black holes are an obvious candidate[7] for this dark matter.

[7] The most recent explanation is that of 'brown dwarfs', sometimes rather slightingly called 'failed stars', i.e. bodies a hundredth of a solar mass which are not very luminous because they are too small for thermonuclear reactions to occur in their cores. Observational programmes on microlensing are essentially aimed at discovering them.

However, various observational constraints rule out too great a population of giant black holes.[8] If, for example, black holes much greater than a million $M_\odot$ exist in the halo of spiral galaxies – that is, outside the bulge and disc where most of the visible matter is assembled, (see Chapter 17) – their presence would manifest itself in at least two ways. First, they would multiply the images of distant stars by acting as gravitational lenses, and secondly they would cause the galactic discs to thicken by increasing the velocity of the stars found there. Neither of these phenomena have been observed.

On the other hand, primordial black holes of a million $M_\odot$ are not ruled out. Most of the galaxies, giant or otherwise could hide giant black holes at their centre, which would probably be primordial in origin, formed very early in the history of the Universe. It may be that black holes are the seeds which made possible the subsequent formation of galaxies.

[8] As will be shown in Chapter 17, it is possible that all the galaxies have a very massive black hole at their centre; but to resolve the problem of the missing mass, there would also have to be many giant black holes located outside galactic nuclei.

16

The zoo of X-ray stars

Too large for its mass to evaporate by thermal radiation, too small to bend the light of distant stars, an *isolated* stellar black hole is condemned to invisibility.

However, a black hole is never completely isolated. It inhabits the interstellar medium, swallowing the surrounding matter and feeding itself. A black hole which consumes matter always leaves behind some remains: the swallowed matter signals its disappearance by the emission of electromagnetic radiation. However, interstellar gas is too tenuous to produce much luminosity. A 10 $M_{\odot}$ black hole slowly consuming the surrounding gas would have the pale appearance of an isolated white dwarf and would at best be detectable only at a distance of a few light years. But even if there were a billion black holes in the Galaxy the closest of them would probably be over 100 light years away.

What is left for the astronomer who wishes to detect a black hole? The answer is *binary systems*. Single stars are in the minority and as a stellar residue a black hole is no different. Many of them should therefore belong to binary systems. But black holes are disguised even more than their compact cousins, white dwarfs and neutron stars, when they belong to a binary system. They do have a kind of signature, however, and decoding this has been one of the most fruitful paths followed by astrophysicists over the past 20 years.

The spectre of a shared life

Black hole or not, binary stars are rarely viewed directly; in most cases only a single component is visible with a telescope.

How can an astrophysicist detect whether a star has a partner or not?

Gravitation holds the key to the problem. In a binary system, the partners rotate about the common centre of gravity obeying the laws of celestial mechanics. In certain pairs of stars near the Solar System, astronomers are able to observe the two partners executing their slow elliptical dance. However, more often than not the orbital motion can be identified only by using the very accurate techniques of *spectroscopy*.

Like the Sun, visible light from the stars is a mix of the colours of the rainbow, from the long (red) wavelengths to the short (violet) wavelengths. A spectrograph is a device which, like a prism, splits up the visible light of a star into its different colours on a screen. Stellar spectra appear as a continuous gradation of colours on which are superimposed very narrow dark lines, called *absorption lines*. The existence of an absorption line shows that light at that frequency has a decreased intensity. What causes this?

The atmosphere of a star is composed of certain atoms: hydrogen, helium, carbon, oxygen, calcium, and so on. Each atom is able to absorb light of certain characteristic wavelengths. More precisely, the electrons orbiting the nucleus capture the energy of some of the incident photons and jump to higher energy levels. The light from the hot cores of stars reaches astronomers only after passing through the 'filter' of intermediary atoms, where some of its energy at certain wavelengths is lost.

It is possible to produce a 'reference spectrum' in the laboratory for any atom. The absorption lines of a stellar spectrum are a kind of 'signature' which when compared with the reference spectrum reveals the chemical composition of the star's outer layers and provides information on the surface temperature, size, intrinsic luminosity, and so on.

There is a class of binary stars in which a single component is visible to a telescope, but whose spectra reveal lines which oscillate periodically around a mean position. They are called *spectroscopic binaries*, because the displacement of the spectral lines indicates the motion of the star around an unseen companion.

A certain shift

The apparent shift in electromagnetic frequencies caused by the motion of a light source with respect to the receiver is called the *Doppler effect*:[1] as a source approaches or recedes the received frequencies are increased or decreased with respect to the emitted frequencies; this shift is larger for higher velocities (Figure 57).

An amusing illustration of the Doppler effect is the story of a driver who was taken to court after jumping a red traffic light. Thinking he was very clever, he explained that the reason he jumped the light was that because of the speed of his car the red light appeared green. The judge, remembering his physics, calculated that the man must have been travelling at 100 000 km/s for the red light to have appeared green via the Doppler effect. Smiling at the motorist he said, 'I accept your argument, you are booked for speeding!'

The Doppler effect is also used in police speed trap radars, the bane of motorists in a hurry. It also has many useful applications in astronomy. An astronomer is something like a blind man with very keen hearing, who knows the reference frequency of fire engine sirens and can estimate the velocity and direction in which they are travelling. Astronomers measure the motion of stars by 'listening' to their light using a spectrograph. This process is particularly good at revealing the binary nature of stars without a visible companion.

It is obvious that in a binary system, the orbital motion of a visible star around the common centre of gravity is revealed by the fact that it appears alternately to approach and to recede from the astronomer, while its dark companion does the opposite.[2] The frequency of the received radiation must therefore increase (become 'more blue') during the approach phase and decrease (become 'more red') during the recession phase. The shift affects

[1] The Austrian physicist Christian Doppler discovered this effect in acoustic waves in 1842, but it was the French physicist A. Fizeau who generalised it to include light waves.

[2] Except in the unlikely situation where the Earth is located exactly perpendicular to the orbital plane of the system.

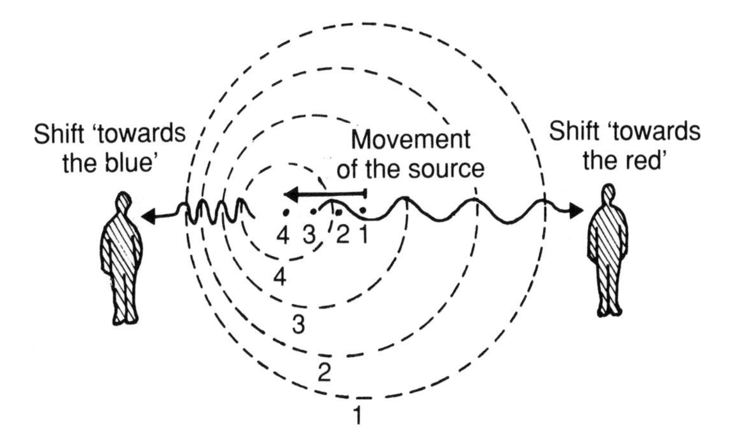

Figure 57. The Doppler effect.

A siren emits spherical sound waves which expand at the velocity of sound. When the siren is moving with respect to stationary observers, the circles tend to get closer together as it approaches. The apparent number of oscillations per second, i.e. the received frequency, increases, and the wavelength decreases: the pitch of the siren becomes higher. Similarly, as the siren recedes from the observer, the sound waves it emits appear to stretch out and the observer hears a lower pitch. The siren can be replaced by a light source, and the Doppler effect alters electromagnetic frequencies in a similar fashion.

the whole spectrum, and the absorption lines move *en bloc* to the red or the blue, oscillating between two extreme positions (Figure 58). This behaviour is the signature of a spectroscopic binary.

Once the astronomer is sure the star is a binary, he can try to discover the nature of the unseen star. An invisible star is not necessarily a black hole: far from it in fact. It could be one of the many stars with very small masses whose luminosity is too faint to be seen – either because the star is very distant or because it is drowned out by the brightness of its companion, like a firefly which becomes invisible as it flies near a lamp.

The obscure star could also be a star of average mass which has gravitationally collapsed. The list of stellar residues comprises white dwarfs, neutron stars and black holes. One might think that the trademark of a black hole is just a question of mass; neutron stars and white dwarfs cannot be greater than 2 or 3 $M_{\odot}$. However, there are many pitfalls in trying to identify a black hole in a binary system. A hot, bright, massive star can remain obstinately hidden because it is surrounded by dust which obscures it. There is a case exactly like this: Epsilon Aurigae is a spectroscopic binary whose unseen companion has a mass of about 8 $M_{\odot}$. This is much greater than the allowed mass of a white dwarf or neutron star. However, the visible component is eclipsed every 27 years and the eclipse lasts 2 years. A black hole would be too small (25 kilometres in radius) to cause eclipses of such long duration. Epsilon Aurigae's companion is just a large star hidden by dust.

Fortunately, a black hole hunter can use other signs. In particular he knows that stars in a binary system evolve differently from single stars, particularly when one of them is gravitationally condensed. Isolated, a stellar residue with a small surface area is invisible for most of the time (except for radio pulsars). In a binary it is transformed. In the case of white dwarfs the binary system is the site of striking events, as in cataclysmic variables and novae (see Chapter 5). For neutron stars and black holes the transformation is even more spectacular. It produces a veritable zoo of high energy astronomical phenomena whose common characteristic is that they are all visible in the X-ray region. The development of X-ray astronomy at the beginning of the 1970s revolutionised the accepted picture of the Universe.

Flying observatories

X-ray astronomy could only begin within the space age. X-rays are absorbed by the atmosphere and so can only be measured by detectors in space. An X-ray detector is much smaller and less visually impressive than an optical telescope, which uses mirrors to reflect and amplify light. The energy transported by X-ray photons (and to an even greater extent by gamma-ray photons) is so great that instead of being reflected by a normal mirror, the photons penetrate it and are lost. So astronomers use special detectors to capture X-rays, which respond to the effects

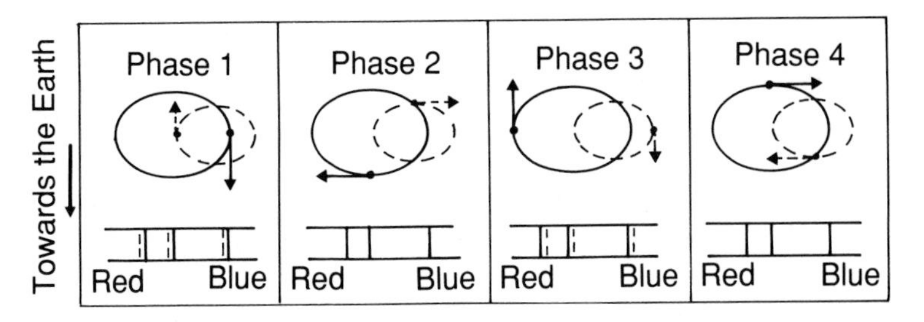

Figure 58. A spectroscopic binary.

The spectrum of a visible star (orbit shown by a continuous curve) oscillates periodically from one side to the other of an average position, revealing the presence of a companion (orbit shown by a dashed curve).

of high-energy photons as they pass through charged metal or gas.[3]

The earliest instruments were flown on rockets and balloons. Many X-ray sources were discovered and named after the constellation in which they were found (for example Scorpius X-1 was the first X-ray source discovered in the constellation of Scorpio). The relatively peaceful image of the Universe which astronomers had built up from observations in the visible and radio frequencies began to crumble away. Once artificial satellites had been constructed which could monitor the 'X-ray sky' over very long periods of time, that image was completely overturned. Suddenly, from all parts of the Universe, sources as diverse as stars, galaxies and galaxy clusters were found to emit a profusion of electromagnetic radiation carrying 100 to 100 million times the energy of visible light.

Rockets had some advantages over satellites because they were cheaper and much quicker to put into operation. It takes only a few months to complete a rocket project, whereas quite a few years elapse between proposing a satellite project to a funding organisation and the actual launch. However, rockets fall back to Earth very quickly and allow only a few minutes' observing time. In the

[3] The famous 'Geiger counters' which measure radiation levels on the Earth's surface operate on a similar principle.

entire rocket era the sky was observed for barely an hour, whereas a single satellite can function for several years.

Freedom

The astronomical community dreamed of an X-ray satellite which could examine the sky 24 hours a day. This dream was realised thanks largely to the effort by Riccardo Giacconi and his team at Harvard University. On 12 December 1970, the 42nd satellite in the 'Explorer' series was launched from a platform in the Indian Ocean off the coast of Kenya, into an equatorial orbit. The satellite was called *Uhuru*, which means 'Freedom' in Swahili, in commemoration of the seventh anniversary of Kenya's independence.

Of the many X-ray satellites, *Uhuru* was one of the brightest jewels because it was the first to make an accurate X-ray map of the heavens. A single X-ray detector can locate a point source with only mediocre accuracy. To overcome this, *Uhuru* carried *two* detectors placed back to back, which swept the whole sky little by little as the satellite slowly rotated. Each time an X-ray source came into their line of sight, a signal was transmitted to the Earth, and knowing the orientation of the satellite, the direction of the source could be determined with much greater accuracy, assigning it to a small 'error box'. *Uhuru* functioned until its batteries ran out in spring 1973, by which time it had located nearly 350 new X-ray sources.

After *Uhuru* a number of other satellites were devoted to the study of the X-ray sky, including the High Energy Astronomical Observatory (HEAO) series. Of this series, the second satellite produced the most spectacular results; it was called *Einstein*, to commemorate the centenary of the birth of the man who in his own way opened new windows on the sky: those of the mind. Europe has been well represented in space astronomy. In the high energy region which concerns us here, we can mention the Soviet satellite *Granat*, launched in 1990, devoted to 'hard' (high energy) X-rays and 'soft' gamma rays. The results are already prolific and still increasing, and a consolation to astronomers for the problems of the US Space Telescope in the optical domain.

X-ray pulsars

More than half of the X-ray sources found by the satellites belong to our Galaxy; the others are the nuclei of active galaxies or the very hot gases of galaxy clusters. Of the galactic sources, the majority are associated with various forms of collapsed star: supernova remnants expanding into interstellar space, white dwarfs, and more importantly, binary systems containing a neutron star.

At the beginning of 1971, *Uhuru* detected Centaurus X-3, a variable X-ray source whose average luminosity was 10 000 times greater than the Sun emitted at all wavelengths. In addition, the radiation from Centaurus X-3 emitted regular pulses every 4.84 seconds. Such a short period suggested that the source was a rapidly rotating neutron star, like a radio pulsar. However, Centaurus X-3 was different, because every 2.087 days its X-ray emission stopped for almost 12 hours. This meant that the source was part of an eclipsing binary system; its occultation corresponded to the phase where the pulsar passed behind a giant companion. A new and fruitful branch of astronomy began, *the study of binary X-ray sources*.

Following the discovery of Centaurus X-3, other 'X-ray pulsars' were soon discovered. One of the most interesting of these is the Hercules X-1, which varies periodically every 1.24 seconds and whose binary nature has been confirmed by several independent means. First and foremost, its X-ray emission is occulted for 6 hours every 1.7 days. In addition, extremely accurate measurements of the arrival times of the X-ray emissions show regular fluctuations around the average period of 1.24 seconds, the shift being caused by the orbital motion around a companion, with a period agreeing exactly with that of the eclipses. To confirm this, very fine optical measurements of Hercules X-1 were made which revealed the companion in the visible region. It was a star whose luminosity was also eclipsed every 1.7 days. Hercules X-1 was a spectroscopic binary discovered 'in reverse', since the compact component was discovered first through its X-ray emission and then used to locate the 'normal' optical component.

What is the mechanism which causes X-rays to be emitted from binary sources? An important clue comes from the fact that these binaries all have short orbital periods, indicating that the distance between the partners is very small. This closeness enables the neutron star to capture gas from its companion by using a sort of 'gravitational vacuum cleaner' which works in the following way. Around a single star there are points where the gravitational field has a constant value; these 'contours' form spheres centred on the star. If we look for the same points in a binary system, the surfaces are more complicated (Figure 59). One of them represents a surface of gravitational neutrality between the two partners. It has a figure-of-eight shape, each loop surrounding a star, and is called the *Roche lobe* after Edouard Roche the French mathematician from the University of Montpellier who first studied the problem in the 1850s. Compact stars like neutron stars are simply point sources within the Roche lobe. On the other hand non-collapsed stars may occupy most of the lobe, or even more than it, as in the case of a red giant. The X-ray pulsars such as Centaurus X-3 and Hercules X-1 can be explained by a binary system in which one of the components is a neutron star and the other a giant star which fills its Roche lobe. The latter easily loses material, mainly at the point of contact between the two lobes. As it passes from one lobe to the other, the gas comes under the control of the neutron star. For Centaurus X-3 we can estimate that a mass equivalent to that of the Moon is transferred each year from the giant to the compact star.

As in a radio pulsar, the neutron star of an X-ray pulsar can spin rapidly and possesses an enormous magnetic field, inclined to its rotational axis. In this case the gas from the companion star does not fall directly onto the neutron star, but is slowly dragged by centrifugal forces into a 'spiral' movement. The gas therefore forms a fairly thin *accretion disc*, which is disrupted at a distance from the neutron star where the magnetic field has a greater energy than the rotational energy of the gas. At this point the gas is sucked from the accretion disc and channelled by the magnetic field lines towards the magnetic poles.

The X-ray emission is caused by the impact of the gas on the star's solid crust. It is easy to understand how a gravitational field is able to convert the energy into radiation if we consider the idea of hydroelectricity. As water falls from a sufficient height, it converts its potential energy into kinetic energy; it is travelling very

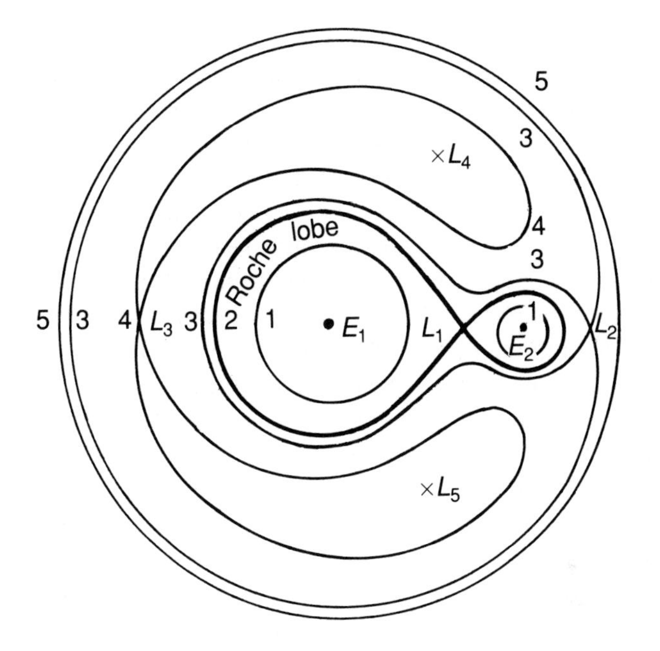

Figure 59. The gravitational field of a binary star system.

Each solid curve, called an 'equipotential', traces the set of points where the so-called 'Roche potential' has a constant given value. This Roche potential describes the contributions of the gravitational fields of the two stars E_1 and E_2 and of the centrifugal forces in the binary. The points indicated, called Lagrange points, are the positions of stable equilibrium (for L_4 and L_5) and unstable equilibrium (for L_1, L_2 and L_3). Near to each of the stars, the equipotential surfaces are spherical (curve 1) because the pull of the central star is dominant. Far from the stars (curves 3, 4 and 5), the equipotentials surround the whole system and tend towards the spherical again, because at large distance the binary looks like a single star. The intermediate figure of eight (curve 2) is the most interesting feature; it defines two 'Roche lobes', surrounding each star, that join at point L_1, much like a mountain pass between two deep valleys. If a star fills its Roche lobe, its material finds it easy to pass through L_1 into the other lobe: this is the mass transfer between the partners.

fast when it strikes the turbine blades, and its kinetic energy is converted into rotational mechanical energy. By magnetic induction, the mechanical energy is finally converted into electrical energy and radiation. Throughout the process it is the Earth's gravitational field which acts as the motor. A similar process occurs at the surface of the stars. Of course, falling through a given distance, the conversion of gravitational energy to radiation is more effective the greater the gravitational field. A 10 gram ball falling to the Earth's surface liberates only a little heat and infrared radiation. If it fell onto the surface of a white dwarf the gravitational energy released would be much greater and would manifest itself as visible light and ultraviolet radiation. On the surface of a neutron star the gravity is so high that the free-fall velocity would reach 100 000 km/s. Under these conditions, 10 grams of gas crashing onto the surface would release X-rays with an energy equivalent to the bomb dropped on Hiroshima.

In an X-ray pulsar, a trillion tonnes of gas crash onto the magnetic polecaps of the neutron star each second. The polecaps have a diameter of about 1 kilometre, and are heated to a 100 million K, emitting X-rays with a luminosity 10 000 times greater than that emitted by the Sun over all wavelengths. Pulsars proper (i.e. radio pulsars) of course involve the modulation of a radiation beam by the spin of a neutron star.

X-ray bursters

X-ray pulsars are not the only kind of X-ray binary source. In many cases the emission is sporadic rather than regular. Instead of being caused by direct impact on the polar caps, it can emanate from hot spots in the accretion disc and the pulsing need not occur. In addition the companion of a neutron star is not necessarily a massive giant but can also be a dwarf. In this case, the mass transfer can be much reduced (Figure 60). Finally, and more importantly, in the absence of any periodicity, we cannot confirm that the compact star is a neutron star. This category of erratic X-ray sources, similar to the cataclysmic variables, which contain white dwarfs (see Chapter 5), is the place to look for stellar black holes.

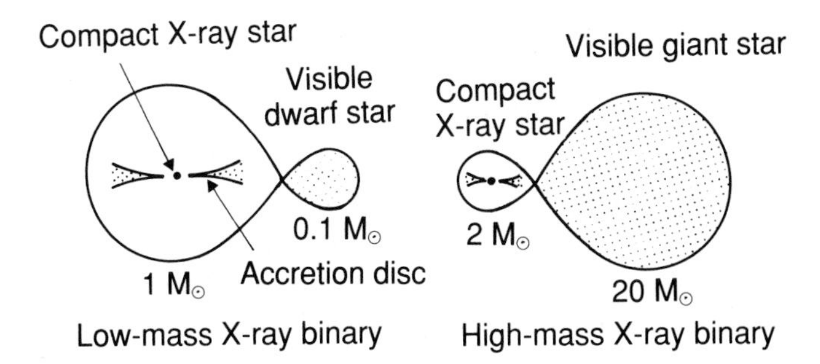

Figure 60. The two types of X-ray binary source.

Since 1975 satellites have been able to detect stars showing violent eruptions of X-rays lasting several seconds. These are called bursters and several dozen of them have been discovered, the majority in our Galaxy. Bursters are similar to novae but they release a lot more energy. They probably occur in close mass-transferring binaries. The difference from novae is that a burster involves a neutron star or a black hole and not a white dwarf.

For a neutron star, the mechanism responsible for the bursters is probably a surface thermonuclear explosion. As in white dwarfs, gravitation plays an important role as a catalyst for nuclear reactions. In neutron stars the colossal gravity causes even more extreme heating than in white dwarfs (where it can cause only the explosive combustion of hydrogen). In the 'quiescent' state, the X-ray source accumulates hydrogen in very hot and dense layers on the neutron star's surface. It is rapidly transformed into helium, but in a non-explosive fashion. The helium covers the star's surface, and when the layer reaches a thickness of a metre explosive fusion is triggered, causing a burst of X-rays. These X-ray eruptions could possibly also be caused by a different mechanism, such as instabilities in the accretion disc. That mechanism does not require the hard surface of a neutron star; a black hole will do just as well.

Some bursters are permanent sources of X-rays (i.e. they continuously emit some level of X-rays), whereas others are visible only in the X-ray band during a burst. On the other hand, like recurrent novae, some bursters explode several times at a frantic pace. There is one ultra rapid burster which has intervals of only a few tens of seconds between explosions.

But recurrence does not mean periodicity. The perfectly regular period of a pulsar comes from the rotation of the neutron star, whereas the recurrence of a burster is caused by repeated detonation of helium accumulated on the surface. We observe that *pulsars never burst and bursters never pulse.* Also, bursters are not necessarily recurrent. This suggests that bursters are found in binary systems older than pulsar binary systems, containing neutron stars so old that they have lost their magnetic field, or in *black hole* systems where it is impossible for matter to accumulate on their surface.

X-ray stars, spectacular as they are, are extremely rare; we can estimate that only one star in a billion emits most of its light as X-rays. There are only about 100 such sources in our Galaxy. This rarity results from the briefness of the X-ray emission phase in a binary system: only 10 000 years in high mass systems,[4] a lightning flash compared with the lifetime of stars (Figure 61). During this time the companion star expands until it exceeds the Roche lobe and becomes so great that it extinguishes the X-ray source.

Gamma-ray bursters

There are even more mysterious sources, whose bursts manifest themselves in the gamma-ray spectrum. These 'gamma-ray bursters' form a completely distinct group from the X-ray bursters: to date there has been no observation which has linked an X-ray burster with a gamma-ray burster.

As is often the case in astronomy, they were discovered by accident. After the treaty forbidding ground-based nuclear experiments of 1963 was signed by the USA and the USSR, the Americans launched a series of military satellites called *Vela* whose mission was to monitor the implementation of the treaty. They were designed to detect gamma rays emitted by secret Soviet bomb tests. To the complete surprise of the American military, an avalanche of data was recorded at Los Alamos! Fortunately for

[4] Low mass systems could live much longer.

world peace, American scientists were able to demonstrate to the armed forces that these gamma bursts originated not on the ground but in space. It was one of the most important astronomical discoveries of the decade.

Since then more than 500 gamma-ray bursters have been detected by a network of monitoring satellites. The gamma-ray bursts last from several milliseconds to tens of seconds. The energies involved correspond to surface temperatures of a billion K; as in X-ray bursters, these temperatures suggest intermittent heating of a neutron star surface caused by the flow of matter.

The main problem with gamma astronomy is the very poor resolution of gamma-ray detectors. This is even worse than for X-ray detectors, where it is already poor. As a result it is very difficult to locate gamma sources and identify them as stars already known to emit at other wavelengths. However, their position can be estimated by combining the observations of several satellites (at least three), which allows 'error boxes' to be defined around them. The main problem with gamma-ray bursters is that within most of the error boxes we do not observe anything unusual. This has led theorists to suggest that gamma-ray bursters could be either isolated neutron stars, or neutron stars associated with very dim dwarf stars. This would explain why gamma-ray bursters are invisible outside bursts. In this model the burst itself would be caused by a surface thermonuclear reaction resulting from the accumulation of gas, but at rates much lower than X-ray bursters. If we recall the mechanisms involved in novae (page 78) and Type I supernovae (page 92), we notice that, paradoxically, the lowest accretion rates onto compact stars produce the most energetic phenomena; Type I supernovae are much more powerful than novae and gamma-ray bursters are more powerful than X-ray bursters.

There is a very famous gamma-ray burster which appears to have been observed in the visible. However, if this identification is correct it raises an enormous theoretical problem. This burster was observed on 5 March 1979 in the vicinity of a supernova remnant located nearly 200 000 light years away. At such a distance the intrinsic energy of the burster is estimated to be a million times greater than other galactic gamma-ray bursters, which is

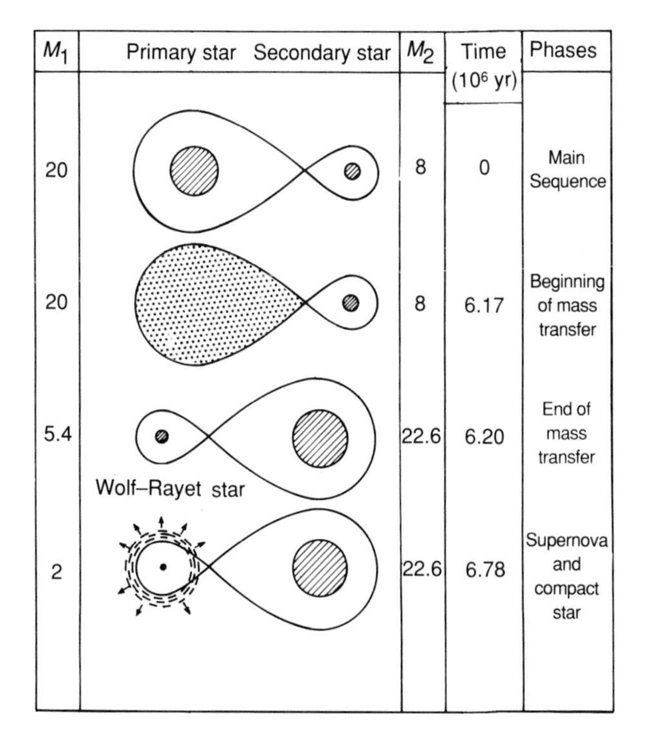

Figure 61. The evolution of a massive binary system.

The diagram shows the evolution of a close pair of stars whose masses are initially 20 $M_\odot$ (primary star, left-hand column) and 8 $M_\odot$ (secondary star, right-hand column). The time is indicated in the second column on the right, in millions of years. The more massive primary star rapidly evolves to the giant stage and fills its Roche lobe after about 6 million years. This is the start of mass transfer to the secondary star. After the transfer phase, lasting 30 000 years, the secondary star has become the more massive, and the primary one is reduced to its very hot helium enriched core. This is a 'Wolf–Rayet star', which explodes as a supernova after about 580 000 years. A

incomprehensible. Either the identification is incorrect and the coincidence of the gamma-ray burster and the supernova remnant is purely accidental (as most of the astrophysicists concerned now believe), or we have to invoke more exotic physical mechanisms than accretion onto a neutron star. Gamma-ray bursters are one of the major puzzles in present day astrophysics.

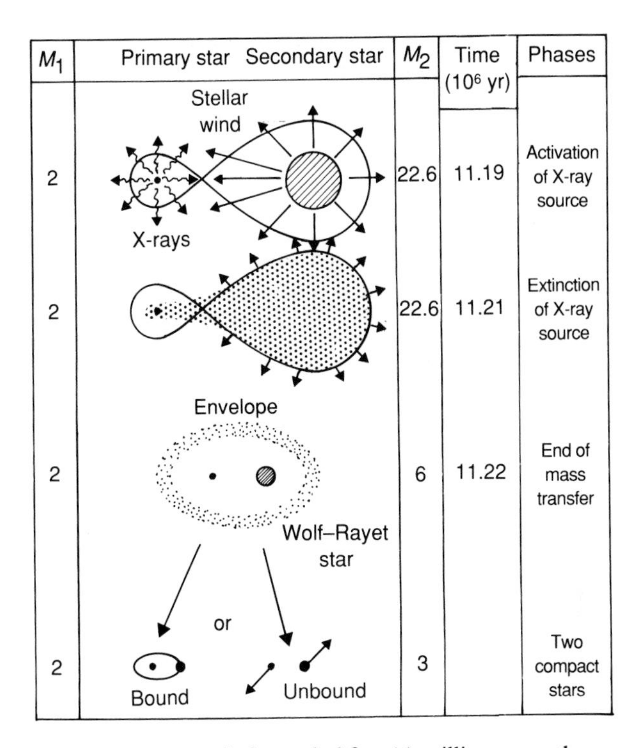

2 $M_\odot$ neutron star is formed. After 11 million years the secondary star has expanded and the stellar wind has blown away much of its matter. Part of the wind material is captured by the compact star, which becomes an X-ray source. This phase lasts only 20 000 years until the secondary star expands beyond its Roche lobe and extinguishes the X-ray source. Mass transfer in the opposite sense begins, leaving a Wolf–Rayet star in close orbit around the neutron star, both of the stars being surrounded by a circumstellar envelope. In the final phase the Wolf–Rayet star explodes as a supernova and leaves another compact star (a neutron star or a black hole). The two stars may remain bound or fly apart.

Looking for a rare animal

Our stroll through the zoo of X-ray stars has taught us that stellar black holes should be sought first in X-ray binary sources that are neither periodic nor recurrent. The first selection process for the candidates consists of measuring the fluctuations of X-ray luminosity over very short periods. Variations in the brightness of any source means that the source is altering its configuration – for example swelling or deforming. Since nothing can travel faster than light, a variable light source can change its global brightness only over a time as short as the time it takes for the light to cross it. Light takes a millisecond to travel 300 kilometres, and so a source fluctuating in less than a millisecond must be extremely compact.

What phenomenon could cause these variations in brightness? Let us use a stellar black hole as an example. It will have a diameter of only a few kilometres, but it is not this parameter which is important for the fluctuations; in fact the source of X-ray radiation is not the impact of matter onto the surface of a black hole, which has only a geometric surface without solidity; rather, it comes from the accretion disc. The inner regions are very hot and turbulent, a little like water when it starts to boil. The disc is locally unstable and 'bubbles' of gas emerge from time to time, are heated for brief instants to several hundred million K and then emit copious bursts of X-ray radiation (Figure 62).

To deduce the characteristic timescale of these fluctuations we have to remember that the accretion disc does not touch the surface of the black hole. There exists a region surrounding the black hole in which circular orbits are forbidden. When the gas in the disc reaches the inner edge, it starts to fall towards the forbidden region and disappears into the black hole so quickly that it does not have time to emit any radiation. We can conclude that these bubbles, which are responsible for the fluctuations in luminosity, can only form at a distance of several Schwarzschild radii from the black hole. They have an extremely short life span once formed; they perform a complete revolution in a millisecond, at a velocity approaching that of light, then dissolve into the surrounding gas and the radiation fluctuation fades. From a distance these variations appear as short bursts in the flux of X-rays.

For several years scientists have been hoping to see this type of ultra-rapid variation in the luminosity of an X-ray binary source, which would reveal the presence of a stellar black hole. Circinus X-1, associated with the remains of a 100 000 year old supernova, fluctuates so rapidly that it was considered as a likely candidate. But this was completely wrong. The progress of astronomical

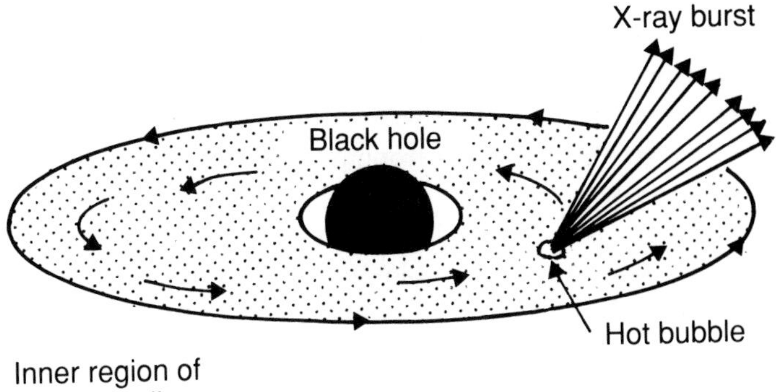

Figure 62. Fluctuations in the accretion disc around a black hole.

instruments showed bursts in Circinus X-1 and other similar sources, proving beyond doubt that they were neutron stars. To discover a black hole in X-ray sources, we must look for something other than feverish activity!

Weighing the stars

'Measure a thousand times and cut once.'

Turkish proverb

The best weapon a black hole hunter has are weighing scales. If we accept the theory of General Relativity and several reasonable hypotheses concerning the state of dense matter, the maximum mass of a stable neutron star cannot exceed 3 $M_{\odot}$. If the 'weight' of a compact star is greater, the only solution offered by modern physics is that it is a black hole.

Unfortunately it is impossible to measure the masses of each of the components of a binary system separately. The astronomer can measure only the spectrum of the optical component – if it is not lost in the spectrum of the accretion disc, as is often the case. The periodic shift of the spectral lines caused by the Doppler effect is linked to the orbital period of the binary system. Astronomers use this to calculate a certain 'mass function' by applying the laws of celestial mechanics. This mass function contains three unknown quantities: the masses of the two components and the inclination of the orbital plane with respect to the direction of observation.

To go further the astronomer has to make several approximations. Spectroscopy of the optical component enables its 'spectral type' to be deduced (see Appendix 1) and taking account of its luminosity enables him to determine its physical parameters: mass, radius, degree of evolution. However, this method of weighing stars using only their spectral type, leads to a large degree of uncertainty.

The other unknown, the inclination of the orbital plane with respect to the direction of observation, is usually unconstrained except in certain binary systems where there are eclipses which enable astronomers to place limits on this inclination.[5]

With these approximations the astronomer can finally deduce the mass of the compact star which interests him. The result he obtains has a certain 'error bar': the middle of the bar is the most likely result and the two extremes correspond to the most 'pessimistic' and most 'optimistic' readings of the data involved. However, since it is a question of proving the existence of black holes the greatest severity is applied, and only candidates for whom the whole error bar is above the 3 $M_{\odot}$ mass limit are considered.

To date only three X-ray sources which satisfy all these black holes criteria have been found.

Cygnus X-1

Cygnus X-1 was discovered in 1965 by an X-ray detector on board a rocket, and observed also by *Uhuru*. In March and April 1971, the satellite registered a rapid variation in X-ray luminosity. By chance, this rapid variation was accompanied by the appearance of a radio source. Radio telescopes are much more accurate than X-ray detectors and the source was located to high precision. It coincided with that of an optically visible star which had long been known as HDE 226 868. This bright star, whose spectral type

[5] The existence or absence of eclipses indicates that the system is seen 'from the side' or 'from above'.

showed it to be a massive hot blue giant of between 25 and 40 $M_\odot$, was incapable of emitting such large quantities of X-ray radiation. Thus HDE 226 868 had to have a compact companion which was extracting gas and heating it to several million K producing the source Cygnus X-1.

To confirm this hypothesis, the spectrum of HDE 226 868 had to be analysed to detect the periodic to-and-fro which characterises the lines of spectroscopic binary systems. The result was convincing; it had an orbital period of 5.6 days, while the maximum shift of the lines allowed the size of the orbit to be calculated. It was extremely narrow: 30 million kilometres. Reducing HDE 226 868 to the size of a football, the companion Cygnus X-1 was like a grain of sand orbiting a few centimetres above its surface.

The absence of eclipses meant that the inclination of the orbital plane with respect to the direction of observation exceeded 55°. Knowing these parameters, one could deduce the mass of Cygnus X-1. The measurements have been repeated regularly over the past 15 years with increasing accuracy. They yield a minimum mass of 7 $M_\odot$, a value much greater than the maximum allowed neutron star. *In Cygnus X-1, astronomers had probably discovered the first stellar black hole*!

Devil's advocates

Although Cygnus X-1 acts as expected for an accreting black hole, other possible explanations have to be investigated.

The relatively 'fragile' part of the reasoning consists of inferring the mass of the optical companion of Cygnus X-1 from its spectral type, then, by subtraction, arriving at the mass of Cygnus X-1. A more careful analysis shows that in fact this operation can be dispensed with and the minimum mass of Cygnus X-1 can be obtained directly, based solely on the absence of eclipses. To calculate the mass in this way we need to know the distance of the X-ray source, which has been estimated at 6000 light years. The minimum mass thus obtained is 3.4 $M_\odot$, which is still sufficiently high to rule out the possibility of its being a neutron star. However, this mass limit decreases with decreasing distance, and we are not very sure of the latter; it would only have to be 10% less for the minimum mass of Cygnus X-1 to fall below the vital 3 $M_\odot$ limit.

Another less serious counter-argument has been proposed, in which Cygnus X-1 is assumed to be a triple system, the visible HDE 226 868 star and two invisible companions. These could be a neutron star and a white dwarf which are so close that they share the same accretion disc. The invisible pair could also be composed of a normal star of about 10 $M_\odot$ hidden by a dust cloud (like Epsilon Aurigae) and a neutron star responsible for the X-ray emission.

There are several important problems with the triple system model. It is difficult to explain how such a configuration could form and survive for a long enough time, since triple systems are very unstable except for a very particular phase of their evolution. However, if Cygnus X-1 was the only candidate black hole the argument would not be conclusive; if there were only one star with strange properties in the Universe, it could statistically be in a very unlikely state. However, this is no longer the case: the abundance of X-ray data gathered over the past 10 years has revealed more binary X-ray sources which are just as likely to be black holes as is Cygnus X-1. The black hole model for Cygnus X-1 and these other systems is in fact the most *conservative* explanation, for it uses the fewest arbitrary hypotheses. It therefore obeys the main rule of scientific methodology: *Occam's Razor* (see page 174). There is little doubt that the number of stellar black holes observed will increase over the next few years.

The Gang of Three

Those put off by numbers can refer directly to Figure 63, an identikit picture of the likely members of the Gang of Three, the select band of stellar black holes. However, it is not without interest to describe some of their characteristics.

One of them, called LMC X-3, is not in our Galaxy but in the Large Magellanic Cloud (LMC), one of the two nearby galaxies which are visible with the naked eye in the southern hemisphere named after the famous Portuguese explorer who first recorded them in his log. The optical partner of LMC X-3 is a hot blue star

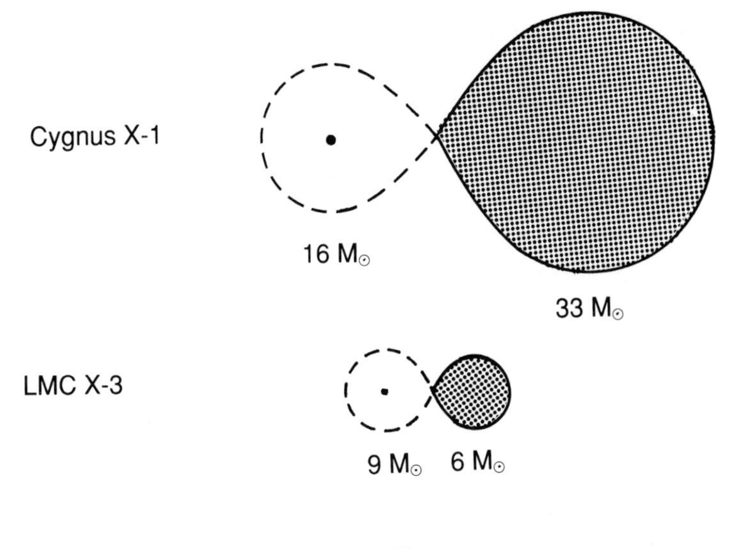

Figure 63. The Gang of stellar black holes.
The relative sizes of the systems are roughly to scale.

estimated from its spectral type to be between 4 and 8 $M_\odot$. The mass of the compact star has been estimated at between 7 and 14 $M_\odot$.

If the same requirements are applied to LMC X-3 as were applied to Cygnus X-1, we ought to forget the spectral type of the companion and deduce a mass depending on the distance. The difference between this case and Cygnus X-1 is that we know the Large Magellanic Cloud is 170 000 light years away. The minimum mass of LMC X-3 is therefore 6 $M_\odot$. So it is even more likely that this star is a black hole than Cygnus X-1.

The third candidate black hole is called A 0620-00. It is about 3000 light years away and is an X-ray source belonging to the subclass of 'low-mass binaries', since the non-compact partner is a dwarf star whose mass is a fraction of a solar mass. There are about 40 low-mass X-ray binary systems in which the non-compact component has been optically identified. However, in most of the cases the X-ray radiation is so intense that it drowns out the optical spectrum, preventing astronomers from determining its orbital parameters and identifying the non-compact component. Fortunately the activity of A 0620-00 in its quiescent state is sufficiently low that it does not erase the visible radiation of its partner. The optical spectrum can therefore be measured. It indeed indicates a spectroscopic binary, whose period is 7.75 hours. From this the mass of A 0620-00 must be at least 3.2 $M_{\odot}$ (if we suppose that the orbital inclination is the least favourable) and probably exceeds 7 $M_{\odot}$.

The remarkable property of the A 0620-00 system is its size, it is so tiny that even 'devil's advocates' could not hide a third star in it (Figure 63). Some astronomers (especially those who discovered it!) believe A 0620-00 to be the *best* candidate for a black hole. It could even be said that A 0620-00 was the first one to be discovered, since it was found on some old photographs taken in 1917 of the Monoceros Constellation, during a burst of optical activity which put it in the category of novae!

The star-galaxy

If we accept that our Galaxy has been producing 1 supernova per century for the last 10 billion years, and that 1 supernova in 100 causes a black hole, then our Galaxy must contain a million stellar black holes. It is therefore perhaps disappointing that our study of X-ray binary sources has so far revealed only 3 possible black holes. It is true that in several other cases there may be black holes, but the error margin is such that we cannot be sure either way. Among the other potential black holes is LMC X-1, also situated in the Large Magellanic Cloud, and half a dozen galactic sources.

Black hole hunters have looked for other ways to test their candidates than mass estimates. Measuring the small-scale fluctuations is one of them, but as we have already seen is not sufficient. Another method has been suggested, based on a 'criterion of similarity': the idea is that if Cygnus X-1 is a black hole, all X-ray sources having a similar X- ray behaviour have a good chance of being one. Cygnus X-1 has a very characteristic spectral 'signature'; it emits its radiation in two states, a 'high' state and a 'low' state. A handful of other binary X-ray sources share the same property, and it is therefore tempting to conclude that they are black holes. However, even here the selection criterion is ambiguous. Of the other members of the Gang of Three, A 0620-00 passes this test but LMC X-3 fails it. Moreover, some of those which have passed the test have already been shown to be neutron stars because of their X-ray bursts, notably Circinus X-1. The fact remains that the best way of testing for a black hole is still to weigh it!

The French telescope Sigma on board the Soviet satellite *Granat* found a bright X- and gamma-ray source in Spring 1990, whose line of sight passes within 300 light years of the Galactic Centre (see Chapter 17). This source has the impossible, but one hopes temporary, name of 1E 1704.7-2942, and is seen by many as the fourth stellar black hole. Moreover Sigma appears to have detected a burst of antimatter from this source, in the form of a massive annihilation of electrons and positrons. According to some high energy astrophysicists, only a black hole can create the required extreme physical conditions in its neighbourhood for such a significant production of positrons.

Finally there are stars which do not belong to the category of X-ray sources but which may still be black holes, although this is difficult to prove. I have already mentioned Cassiopeia A (see page 93), one of the brightest radio sources in the sky and associated with a supernova remnant. The explosion took place in about 1670, but it was not as bright as expected. This supernova remnant does not contain a pulsar or an X-ray source. It is therefore possible that the star which gave birth to Cassiopeia A was very massive and its core collapsed directly into a black hole, preventing the occurrence of a very bright supernova.

One of the most enigmatic stars in the Galaxy is SS 433. It is remarkable not only for the intensity of its spectral lines but also because of the two symmetrical groups of lines, one strongly shifted to the blue and the other towards the red, which oscillate about an average position with a period of 164 days.

Interpretation of the average spectral shift in terms of the Doppler effect indicates that the velocity of the source reaches

78 000 km/s. How could a star move at such a velocity? A clue is given by the fact that these are not absorption lines, coming from stellar light filtering through the envelope, but *emission* lines, emitted by hot gases. These two groups of lines originate in two symmetrical gas *jets* issuing from a central star, each one pointing in turn towards and away from the Earth. Observations in the radio band have confirmed the existence of these jets.

In addition, spectroscopy of SS 433 indicates that it is a binary system harbouring a compact star, either a neutron star or a black hole. The question was debated until 1991, when a reliable measurement of the mass of the compact component was obtained by a European collaboration. It is only 0.8 $M_{\odot}$, which means that the compact star is too light to be a black hole. However, SS 433 is extremely interesting to astronomers because of its extraordinary system of gas jets. Attempts to model this star have stimulated accretion disc theory. In order to understand the origin of the jets, we should remember that a compact star – whether a neutron star or a black hole – cannot accrete arbitrarily large amounts of matter; in fact accretion produces radiation which tends to repel the surrounding matter. The accretion disc functions like a massive star, whose equilibrium is maintained by the pressure of radiation issuing from its thermonuclear core. In the accretion disc the equilibrium is maintained in a similar fashion by the opposing forces of gravitation and radiation pressure.

What happens if the companion star which supplies the gas expands beyond the Roche lobe and starts to dump such large quantities of gas that the compact star cannot take it all in? The excess must be ejected. It is clear that the pressure of gas which accumulates in the disc encounters the greatest resistance in the plane of the disc, where the new gas is continuously arriving. Consequently the path of least resistance is the direction perpendicular to the plane of the disc. The compact star relieves itself of the excess by ejecting the surplus gas in this direction. The two powerful jets of hot gas ejected from SS 433 probably show this mechanism at work (Figure 64).

SS 433 is a fascinating model of what happens on a much larger scale in the heart of active galaxies and quasars: a pair of ultra-rapid jets bursting from a compact source. In this case the mass involved

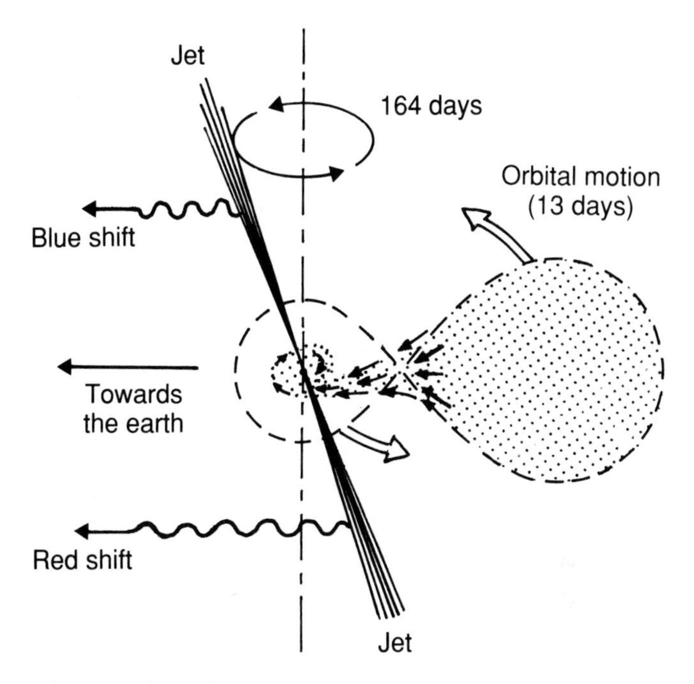

Figure 64. The SS 433 system.

Gas from the star filling its Roche lobe is captured into the accretion disc surrounding the compact companion – either a neutron star or a black hole. Jets of gas are ejected from the two faces of the disc, perpendicular to its plane. As the disc is inclined with respect to the orbital plane of the binary, the two jets precess around the axis perpendicular to the orbital plane, with a period of 164 days. Each of the jets approaches and retreats in turn from the terrestrial observer, producing via the Doppler effect a spectral shift which alternates between the blue and the red.

is not 3 or 10 $M_\odot$, but 10 million or a billion $M_\odot$. Such masses could not possibly be those of a neutron star. This could be the realm of *giant black holes*. These are discussed in Chapter 17.

17

Giant black holes

When a black hole is produced as a result of the gravitational collapse of a star it may have a maximum mass of about 10 times that of the Sun. However, the theory of gravitational collapse allows us to imagine black holes of a thousand, a million or even several billion $M_\odot$ (see Appendix 2). What mechanisms could produce giant black holes?

Three are known. The first of them (already been mentioned in Chapter 15) is that giant black holes could have developed as a result of the condensation of lumps in the early Universe. The second mechanism appeals to the tendency towards irreversible growth that is one of the characteristic properties of black holes;[1] provided that the surrounding environment is sufficiently rich in matter a giant black hole could result from an initial 'stellar seed' of 10 $M_\odot$ produced during a supernova. The third mechanism is the direct formation of a massive black hole by the gravitational collapse of a *star cluster*.

Apart from the possible primordial origin, the formation of giant black holes requires an enormous amount of matter in the form of stars or interstellar gas, confined within a small enough region for its evolution to be controlled by gravity.

In the Universe matter is much more concentrated in galaxies than in intergalactic space,[2] and inside galaxies the greatest

[1] In the present context, the evaporation of quantum mini black holes is of course negligible.

[2] At least the luminous matter.

concentration of matter is at the centre. If there are giant black holes then this is where we should start looking; we shall begin with our own galaxy.

Identik picture

Milky Way, O shining sister
Of the white streams of Canaan
And the white bodies of women in love,
Shall we follow your track towards other nebulae,
Panting like dead swimmers?

Guillaume Apollinaire

The Milky Way is a disc 100 000 light years across and 300 light years thick. This is exactly the same as the ratio between the diameter and the thickness of an LP record. At the centre there is a large swelling, the bulge. The disc and bulge are located in a much more tenuous sphere of stars called the halo (Figure 65).

There are about 100 billion stars in the Milky Way. The majority are found in the disc. The Sun is situated on the periphery, about 30 000 light years from the Galactic Centre. Besides stars the disc is also filled with gas and dust. The distribution of matter is not at all uniform; it is much denser in the *spiral arms* which give the Galaxy its characteristic shape.

The disc is continuously subjected to dynamical and chemical transformations. The arms rotate and deform, and stars are born in the large molecular hydrogen clouds; the larger ones explode rapidly as supernovae and sow the surrounding region with complex chemical elements which will be incorporated in subsequent generations of stars. The halo on the other hand is practically dead, a vestige of the primitive Galaxy. Stripped of gas, it contains only very old stars which were probably formed along with the Galaxy, 15 billion years ago. All the massive stars exploded a long time ago, leaving neutron stars and probably black holes. The medium-sized stars have left the Main Sequence and some have already produced white dwarfs; others are still subject to great upheavals; they are pulsating red giants, variable and very bright.

Finally there are many low mass stars in the halo which are carefully consuming their hydrogen and will live for a long time to come.

The most important characteristic of the halo is not the nature of the stars which populate it but the way they gather together in *globular clusters*.

Globular clusters

Unlike 'open' clusters, the loose groups of young stars mostly found in the disc, globular clusters are found throughout the galactic sphere. Each cluster contains several hundred thousand stars within a diameter of less than 150 light years. They appear as solid spherical balls bound by gravity. The most famous one is in the constellation of Hercules and, although visible to the naked eye, it takes a powerful telescope to resolve the luminous ball into individual stars. At the centre of this cluster, the star density is 20 000 times greater than it is in our own vicinity of the Galaxy. If a planet orbiting one of these stars was inhabited by astronomers, the heavens they studied would be truly magical. Night would be almost unknown because the sky would always be brighter than ours at full moon. These astronomers would know much about stars but next to nothing about distant galaxies, because their weak light signals would be drowned out by the nearby stars.

The globular clusters which accumulate so many stars within such a small volume are very bright because they contain variable giants. Because they are so luminous, these globular clusters can be used to define the frontiers of the Galaxy.[3]

The distribution of globular clusters also enables astronomers to measure the *dynamical centre* of the Galaxy. They follow very eccentric orbits, with the centre of the Galaxy at one of the foci. It takes about 200 million years to complete one revolution and the globular clusters frequently cross the galactic disc. Each time they traverse it powerful tidal forces strip away the peripheral stars which are least strongly bound to the cluster.

Because the globular clusters are so compact the details of their

[3] There are globular clusters in most other galaxies, regardless of their type.

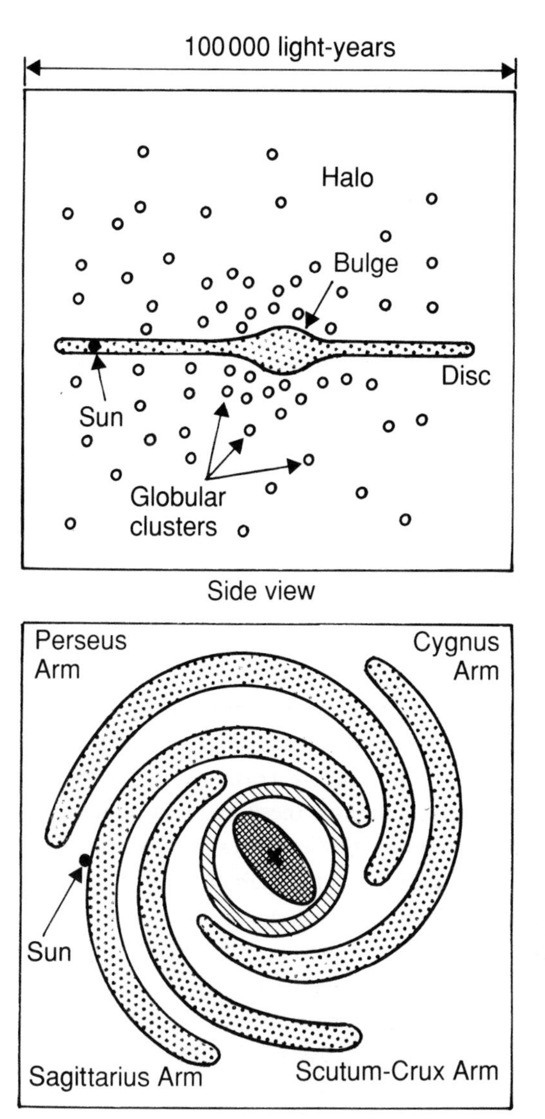

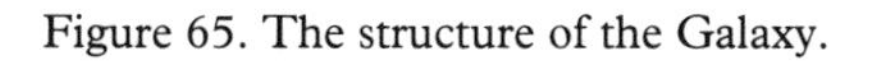

Figure 65. The structure of the Galaxy.

complex evolution are not yet fully understood. We do not know if the formation of a large central black hole occurs as a result of the fusion of stars. The broad outlines of their evolution can, however, be sketched as follows.

It is known that globular clusters evaporate. Just as stars radiate their energy as heat and light, star clusters lose energy by ejecting entire stars. The reason for this is quite simple: as the stars skim past each other they accelerate – the smaller stars accelerate more than the larger ones – and some of them acquire enough speed to break free of the gravitational pull of the star cluster. The galactic halo may be just the 'steam' of globular clusters.

To compensate for this, the remaining stars – the largest ones – draw together and the cluster contracts. Clusters are radically different from stars because stars begin thermonuclear reactions which counteract the gravitational contraction and stabilise the system. In clusters, though, the contraction energy is converted into kinetic energy which further increases the velocity of the stars. More and more of them acquire enough energy to escape. The evaporation speeds up and contraction of the core increases: the cluster is unstable. The evaporation of globular clusters is reminiscent of that of mini black holes, which is not surprising. They both share the thermodynamic properties of purely gravitational systems (already mentioned in Chapter 14), their temperature[4] increases as they lose energy. In globular clusters this instability causes the gravitational collapse of the nucleus, which is called the *gravothermal catastrophe*.

Astrophysicists have therefore wondered if the nuclei of globular clusters do not favour the formation of massive black holes, of several hundred or several thousand $M_{\odot}$, resulting from the merging of massive stars at the bottom of a central gravitational well. This theoretical hypothesis is supported by some observational arguments. If a giant black hole lies at the centre of a globular cluster, the stars sucked into the gravitational well should accumulate in orbits bound to the black hole and reinforce its

[4] This is the temperature deduced from the average agitation velocity of the stars, in the same way that the temperature of a gas is linked to the average agitation velocity of the molecules.

central luminosity. Several evolved globular clusters do show this central 'cusp' of luminosity. Further, about 10 globular clusters are also X-ray sources. This is a considerable proportion compared to the total number of stars contained in globular clusters. There are about 50 X-ray sources spread throughout the Galaxy. The total mass of the Galaxy is 100 billion $M_\odot$, and the combined mass of the globular clusters is 2000 times less. If the ratio between the number of X-ray sources and the number of stars in a globular cluster was the same as in the Galaxy, it would be very improbable to find a single one in any of the globular clusters. The fact that there are 10 shows that globular clusters are highly favourable sites for X-ray sources. These are usually associated with compact stars, capable of capturing the surrounding gas and heating it up to several million K. A 1000 $M_\odot$ black hole could do this by consuming the gas lost by nearby stars.

However, the situation is not as favourable to black holes as at first appears. Recent observational and theoretical progress has tended to rule out the existence of massive black holes in globular clusters. The reasons for this are as follows. If the dynamical evolution of a globular cluster always ended in the production of a massive black hole, then a high proportion of globular clusters should have the central luminosity peak. This property is only rarely observed. There must then be a mechanism which is able to interrupt the gravitational collapse of the nucleus and stabilise it at a 'normal' size. This is the *formation of binary systems*.

This is an entirely natural explanation, but we had to wait for complex numerical calculations on powerful computers to be convinced that encounters of many stars confined in a small space inevitably favour the production of binary systems! Celestial mechanics requires that when a massive binary forms at the centre of a globular cluster, each star which approaches too close will be ejected by a gravitational kick to a more distant orbit. This is the basic reason why the 'forced' creation of a binary system can interrupt the contraction of the nucleus. The fact that a number of globular clusters are also X-ray sources is additional evidence for the presence of binary stars. Indeed, since the X-ray sources in globular clusters are not much brighter than those in the galactic disc, there is no reason to invoke mechanisms other than the accretion of gas onto a neutron star or black hole in a close binary system (see Chapter 16). Besides, the X-ray sources of globular clusters are frequently bursters, which have outbursts lasting several seconds. This phenomenon is frequently associated with compact stars living with a partner.

Finally, improvements in the resolution of X-ray detectors have shown that the globular X-ray sources are slightly displaced from the centres of globular clusters. A giant black hole would dominate the stellar dynamics of the core of a globular cluster and would therefore occupy a central position.

The hypothesis of massive black holes at the centres of globular clusters has thus recently 'fallen into disrepute'. However, of the several hundred globular clusters in our Galaxy and the 15 000 in the large elliptical galaxy Messier 87, the chances that a sufficiently massive cluster has developed a central black hole is certainly not negligible.

Sagittarius or the galactic black hole

The dynamic centre of the Galaxy, in the direction of Sagittarius, is hidden from astronomers by large clouds of gas and cosmic dust. Only one out of every trillion photons emitted in the visible spectrum survives the 30 000 light year journey to the Earth. Traditional telescopes are not much use under these conditions. Fortunately for astronomers, electromagnetic radiation has a large spectrum, ranging from radio waves to gamma rays. Of this spectrum, radio waves, infrared and X-ray radiation are not affected by the dust clouds. So the Galactic Centre can be studied using radio telescopes and satellites.

The Galactic Centre has a diameter of 30 light years. The 'bolometric' luminosity (the sum of all the contributions of all wavelengths) is 10 million times that of the Sun's luminosity. There are two radio sources within this region. One of them, Sagittarius A East, has all the characteristics of a supernova remnant. The other, Sagittarius A West, is the superposition of two types of radio emission: one is 'thermal' and comes from natural radiation of a cloud of hot gas; the other comes from the heart of Sagittarius A

West itself and is not thermal but is produced by electrons forced to move at velocities close to that of light.[5]

This 'non-thermal' radio source, called Sagittarius A*, is the most powerful radio source in the Galaxy. Its luminosity is 10 times greater than the Sun's optical luminosity. However, the most remarkable thing about it is its compactness: the emission comes from a region smaller than 3 billion kilometres, similar in size to the orbit of Saturn or the diameter of a red giant. It is impossible to put a star cluster in such a small volume. The radio emission is therefore from a *single* source. Now only a few kinds of source are capable of emitting radio waves: pulsars, the remains of supernovae, binary X-ray sources, and massive black holes. Let us consider each in turn.

It cannot be a pulsar: the brightest known pulsar has a luminosity which is 10 000 times less than Sagittarius A*, and in any case the radio emission from the centre of the Galaxy does not pulse, it is remarkably stable.

It cannot be a binary X-ray source. The luminosity of binary X-ray sources fluctuates at all wavelengths; their average radio luminosity is 100 000 times less than that of Sagittarius A*, and reaches a tenth of it at the maximum of a burst. Further, while the radio luminosity of Sagittarius A* is too great for it to be a close binary system, its X-ray luminosity is far too weak.

A recently-exploded supernova remnant would be a powerful radio source. The problem with this explanation is the expansion velocity: it would be much greater than the 15 km/s observed in Sagittarius A*.

It is impossible that the source responsible for the radio emission has an ordinary stellar mass. If this were the case the star would have a velocity which is typical of those stars in the centre of the Galaxy, 150 km/s. This velocity would be observed as a slow motion of the radio source on the celestial sphere, and this has not been detected. Measurements confirm that the source is at rest at the centre of the Galaxy; its mass must therefore be greater than that of a star. The hypothesis of a black hole of several million $M_\odot$ in a state of slow accretion is the only model compatible with all the

[5] This is synchrotron radiation; (see also page 269).

radioastronomical observations. This hypothesis must now be tested by observations through another of the 'windows' on the Galactic Centre: the infrared.

Infrared astronomy has been studied for only a few years, since sophisticated detectors were placed on board satellites such as IRAS.[6] It was discovered that the compact radio source of Sagittarius A* coincided almost exactly with an infrared source called IRS 16. This is a very compact source probably responsible for all the luminosity in the surrounding 30 light years, and heats and illuminates the gas of Sagittarius A West. What is the nature of IRS 16?

Stars in their red giant phase are powerful infrared emitters. By measuring the infrared flux from IRS 16 it is possible to trace it back to the red giants causing it. If we assume a 'normal' proportion of red giants it is possible to deduce the distribution of stars in IRS 16. By this method it turns out that 2 million stars must move within a radius of 5 light years. This extremely high stellar density is 1000 times greater than that of a globular cluster.

However, red giants are not the only source of infrared radiation. Spectroscopic measurements show that gas clouds heated to 300 K orbit around IRS 16 and contribute to its infrared emission. If the red giants can be used to trace the star density, the motion of the clouds is an indicator of the *total mass* of IRS 16. This important information is acquired from the simple hypothesis that the gas moves in circles under gravity. Under these conditions, the orbital velocity of the cloud – as calculated from the Doppler shift – provides a direct measurement of the central mass. These measurements have produced a result which lies between 5 million and 8 million $M_\odot$. Since the total number of stars in this region only add up to 2 million $M_\odot$, there must be an unseen mass of between 3 million and 6 million $M_\odot$. The hypothesis of a giant galactic black hole is therefore strongly supported by infrared astronomy.[7]

We should finally consider the X-ray region. The Franco-Soviet satellite *Sigma* launched in 1990 had the Galactic Centre as a prime

[6] Infra-Red Astronomical Satellite.

[7] On the other hand, X-ray and gamma-ray emission have been detected, indicating the presence of a compact source.

target. The first surprise was the finding of a powerful X-ray source, but one which did not coincide with Sagittarius A* or IRS 16. It is at least 300 light years away. Contrary to what has been said too often, this discovery in no way contradicts the presence of a massive black hole: this will not emit high energy radiation if it does not accrete. As we saw in Chapter 16, the X-ray source found by *Sigma* may in any case be a black hole, but a stellar one in a binary system.

At present many astrophysicists are agreed that the Galactic Centre consists of three structures. First, a 'warm' gas disc with lumps of matter gathered in a 'corona' stretching from 5 to 30 light years from the centre; the inner edge of this ring is strongly heated by a central radiation source. Secondly, there is a cavity within the corona with a 5 light-year radius, containing 2 million solar masses in the form of stars in a very compact cluster. Thirdly, at the centre of this is a slowly accreting black hole of between 3 million and 6 million $M_\odot$.[8]

It is interesting to note that the diameter of a 3 million $M_\odot$ black hole is 20 million kilometres, 100 times smaller than the dimension of the region resolved by present instruments. This resolution will obviously be improved in the next few years, but we should remember that, seen from the Earth, the angular dimension of the galactic black hole is that of a tennis ball placed in an orbit of a million kilometres.

The idea of a large invisible star, similar to those predicted by Laplace, hidden in the centre of the Galaxy was first suggested in 1801 by the German astronomer Johann Seldner. However, Seldner constructed this hypothesis simply to explain the galactic rotation. The mass required was so incredible that Seldner abandoned the idea immediately. The first serious prediction of a giant galactic black hole was in 1971, before there were many radio and infrared data to support it. Suggested by Donald Lynden-Bell and Martin Rees at the University of Cambridge, it was the logical consequence of some previous work by Lynden-Bell, who in 1969

[8] Devil's advocates argue quite reasonably that the motion of the gas clouds could be neither circular nor gravitational but could represent forced ejection by the pressure of radiation from a central star. Under this hypothesis a mass of 300 $M_\odot$ would be sufficient to explain the observed velocities.

suggested that *all* galactic nuclei sheltered giant black holes, whose flow of energy would be regulated by the quantity of gaseous fuel available.[9]

Developments in extragalactic astronomy have tended to support this hypothesis. In comparison with active Seyfert galaxies and above all pulsars, the black hole in Sagittarius is a very poor engine. However, it is likely that in the recent past much more violent events than those which have been observed took place in the heart of the Galaxy. When 2 million stars are bound closely to a giant black hole, there is a chance that every 10 000 years a star will deviate from its circular orbit and fall into the black hole. Once there it would be destroyed by the colossal tidal forces.[10] Part of the debris would be swallowed by the black hole and cause a period of activity which would last several tens of years. The rest of the debris would be thrown into an eccentric orbit. Serious consideration has been given as to whether or not the warm clouds observed in IRS 16 could be the debris of stars that have been broken up by the black hole over the last million years. Their number is compatible with a frequency of one stellar disintegration every 10 000 years.

All this seems to confirm that the centre of our Galaxy is a miniature version of the cataclysmic phenomena which take place at the centres of the most distant galaxies.

The realm of the galaxies

Billions of galaxies are visible to astronomers via modern telescopes. At the beginning of the twentieth century Edwin Hubble, like Buffon before him for animal species, classified the different types of galaxies into groups depending on their morphology: elliptical, spiral, barred spiral or irregular. We have already seen in the Milky Way that spiral galaxies have three components, a bulge, a structured disc, and a diffuse halo. Barred spirals also usually contain two arms wound about a central bar. Irregular galaxies resemble a spiral galaxy without its halo and

[9] In 1964 two Russian astrophysicists, Yacov Zeldovich and Igor Novikov, proposed that the accretion of gas onto a supermassive black hole could be the energy source of the quasars.

[10] The disruption of a star by the tidal forces of a black hole is described later on.

bulge. The elliptical galaxies on the contrary resemble a spiral without a disc but with a large bulge and halo. This type of galaxy includes some of the largest; gigantic swarms of a trillion stars. The main characteristic of elliptical galaxies is that they contain only stars and practically no gas.

It is thought that all the galaxies have similar ages, 15 billion years, and that they have different morphologies because they have different 'metabolisms'. The metabolism of a galaxy is just the rate of conversion of gas into stars, the sign of 'life' in a galaxy. According to this view, elliptical galaxies are those in which the rate of initial conversion was most rapid, and most of the stars were formed before the clouds had time to interact and fall gradually into the disc.[11] Spiral galaxies by contrast are those whose initial metabolism was sufficiently slow that the formation of stars did not take place until after the gas had flattened itself out into a disc. The irregular galaxies are a case of arrested development: less than half the gas has been incorporated in the stars and no precise morphology has appeared.

When viewed solely in terms of metabolism the evolution of the galaxies seems quite peaceful. In elliptical galaxies the metabolism is frozen, while in spiral galaxies there is a slowly diminishing cycle in which stars are born, forge heavy elements during their lifetime, enrich the surrounding gas by exploding, and are then followed by new generations of stars, each new generation assimilating the elements produced by the last

Active nuclei

One of the revolutions of the past 30 years of astronomy has been to understand that galaxies do more than just produce starlight. Some of them shelter a source of intense 'non-thermal' radiation at their centres which is not stellar in origin. The Milky Way is the most obvious of these, although the energy produced by its nucleus represents only three thousandths of the total energy emitted by the 100 billion stars in the disc and halo. However, in

[11] The collision of two gas clouds dissipates a great deal of orbital energy, which tends to make them fall into the 'equatorial plane' of the galaxy.

1% of all observed galaxies, the central activity is so great that the 'non-thermal' activity within a volume as small as the Solar System itself is greater than that of the rest of the host galaxy. These *active galactic nuclei* have very powerful 'engines' at their hearts.

The quasar[12] 3C 273 is an excellent example which illustrates the nature of the problem posed by active galaxies. Of all astronomical phenomena quasars are without doubt the most exciting because of the fabulous amounts of energy they emit. 3C 273 is 3 billion years away and is as bright as 1000 galaxies. It appears to be a point source and so must be very small; measurements have revealed that its diameter is less than a light year. In comparison with the total volume occupied by our Galaxy, 3C 273 is as tiny as the Eiffel tower compared with the Earth. How can it be 1000 times brighter?

The entire problem of active galactic nuclei is shown by this extreme example. What we know about the way the 'central engine' functions in active galactic nuclei today is similar to what we knew about the internal constitution of the stars 60 years ago. We did not know then that the nuclei of stars were fed by thermonuclear reactions. Thanks to developments in nuclear physics we have been able to understand why stars have their observed masses and luminosities, and we can calculate their structure and trace their evolution. For galaxies, no such clear picture has emerged. However, it is probable that the *accretion of matter by a giant black hole* plays a role similar to the liberation of thermonuclear energy in stars. We shall now try to understand why.

Five easy pieces

The family of active galactic nuclei includes many kinds of extragalactic sources, ranging from radiogalaxies to quasars, and including 'Seyfert galaxies', 'Lacertids' and exploding galaxies. To describe in detail the observational peculiarities of each type would go beyond the scope of this book. Our main interest is in their

[12] The term quasar is a contraction of the expression 'quasi-star'; it dates from their discovery at the beginning of the 1960s. Their point-like appearance resembles that of stars; improved instruments have since shown them to be surrounded by nebulae. They are in fact the very bright nuclei of distant galaxies.

common properties, and in particular the puzzle they pose for astronomers: what is the nature of the central engine? There are five pieces to this puzzle: the 'non-thermal' nature of the radiation; the high concentration of mass, the variability in luminosity; the ejection of gas jets to very great distances; and the similarity with normal galaxies.

The active nuclei in galaxies can be observed at nearly all wavelengths: radio, infrared, visible, ultraviolet and X-ray. The *spectrum* is the most remarkable part of this emission, i.e. the distribution of intensity as a function of frequency. It does not look like that of a star or a collection of stars. The radiation emitted by the surface of a star is very similar to that of an ideal 'black body' (see page 210) and is characterised by its temperature; it is called *thermal* radiation. The radiation of active galactic nuclei is, however, *non-thermal*. The most obvious example of this is radiogalaxies of the 'synchrotron' type, i.e. the radiation is emitted by highly energetic electrons travelling at velocities approaching that of light within a magnetic field.

Psychology of the masses

There is a wide range of theoretical and observational arguments suggesting that there are high concentrations of matter in galactic nuclei. The first of them is a very general constraint on the quantity of light that can be emitted by any source, regardless of its nature. A given mass cannot radiate above a certain critical luminosity called the *Eddington limit*. There is a simple explanation for this; in a stable light source[13] the outward force exerted by the pressure of radiation leaving it must not exceed the inward gravitational pressure which holds the source together. The Eddington luminosity is the limiting case where the two forces are equal. If the Sun's luminosity increased 25 000 times it would simply evaporate because it would be unable to maintain the cohesion of its gas; some very hot young giant stars radiate so close to their Eddington limit that they very quickly 'blow away' their gas envelope. If the

[13] It is essential that the source is stable; the luminosity of a supernova is far greater than the Eddington limit!

source consisted of a black hole accreting a gas cloud, rather than a star, the radiation pressure emitted by the gas could not exceed the gravitational force that the black hole exerts on the gas particles; otherwise the particles would be repelled and the accretion would stop.

If we assume that the nuclei of very active galaxies radiate at the Eddington limit, a measure of their luminosity gives an estimate of their mass. The luminosities of active nuclei lie between a 100 billion and 1000 trillion times the Sun's luminosity. Their masses therefore lie between 1 million and 10 billion $M_{\odot}$; the upper limit corresponds to the most active nuclei, the quasars.

The idea of a massive engine is supported by a second theoretical argument concerning *efficiency*. The luminosity of a source always results from the conversion of a certain mass into radiative energy. Let us consider the thermonuclear energy liberated at the centre of stars, which we are accustomed to see as a very efficient mechanism for converting mass into energy. When a kilogram of hydrogen is converted into helium, only 7 grams are dissipated as radiative energy (see page 64); in other words the 'efficiency' of thermonuclear energy is only 0.7%. Let us imagine for the moment that the energy released from an active core was also thermonuclear in origin; this would imply that a quasar has to consume 1000 solar masses each year to produce its luminosity. We have good reason to believe that a quasar shines for several million years, which implies that during its lifetime a quasar consumes the equivalent of an entire galaxy. This seems so excessive, especially when we take into account the small volume involved, that we must ask whether or not a more efficient mechanism than thermonuclear energy is involved.

Now we have already seen with the binary X-ray sources that *energy release in a strong gravitational field* satisfies this requirement. When a kilogram of hydrogen falls slowly into a black hole from an accretion disc, *100 grams* are converted into energy. This is much more efficient than thermonuclear reactions. This simple statement of the formidable energies which can be extracted from gravitational fields has encouraged astrophysicists to look to compact objects for explanations of the most violent celestial phenomena, whether on a stellar scale with novae and X-ray sources, or on a much greater scale with galactic nuclei. A new and vigorous branch of astrophysics called *relativistic astrophysics* developed in the 1970s, concerned with the study of how matter behaves in the gravitational field of a compact object.

Without observational evidence to corroborate them, theoretical suggestions of large compact masses at the centre of active nuclei will remain just suggestions. There are two methods which can be used to measure concentrated masses approximately, but they can only be used on nearby galaxies.[14] The first involves studying the distribution of starlight near the centre. We recall that this method is currently used to examine the centres of globular clusters (see page 258). If a large central mass is present, the stars attracted by the gravitational well will accumulate and increase the luminosity dramatically. The second method consists of deducing the mass from the motion of the surrounding matter; it was successfully used for the Galactic Centre (see page 264). In external galaxies, the velocities of stars near the centre are measured; interpreting these as circular motion around the central mass, we can deduce its value.

These two techniques were used successfully in 1978 to measure the mass of the nucleus of the elliptical galaxy Messier 87, one of the most powerful radio sources in the heavens. The measurements indicate a central mass of between 3 and 5 billion $M_\odot$. In addition, the nucleus of Messier 87 is not bright enough to consist solely of stars. This could be the first observational discovery of a supermassive black hole. However, as in the case of the Galactic Centre, the estimate of the stellar velocities can be reasonably doubted: if the stars were moving radially and not in a circle there would be no large central mass.

The measurements made for Messier 87 were followed by systematic investigation of the nuclei of nearby galaxies. Where there are active nuclei (in Seyfert galaxies) the central masses are usually estimated at between 10 million and 100 million $M_\odot$. The record is presently held by NGC 6240, which appears to harbour a gigantic dark nucleus of 50 billion $M_\odot$. However, the efficiency of

[14] Quasars are excluded from these measurements; their mass is deduced from their luminosity alone.

the 'gravitational engine' requires a mass which is not simply large but also concentrated. In radio galaxies, it is possible to measure the maximum size of the emitting nucleus directly by using *long-baseline interferometry.*[15] Using this technique one can show for the best resolved nearby sources that the central mass is confined within a diameter of less than a light year.

The inconstant heart

Not all very active nuclei are also radio sources. It is possible to determine the sizes of these indirectly from their *variability*.

We have already seen in the Chapter 16 how the variability of a source is an indication of its size, since a change of configuration of the source cannot travel faster than the velocity of light. If, for example, the luminosity of an active nucleus shows significant fluctuations in a day, the source must be confined within a region of one light day, or 26 billion kilometres.

We have also seen that the luminosity of a source can be used to calculate its mass. It is clear that the size of the source has to be greater than the Schwarzschild radius for a black hole of the same mass. A black hole of 100 million $M_\odot$ has the dimension of a light hour, which implies that an active nucleus of 100 million $M_\odot$ cannot vary more rapidly than over an hour. Thus the characteristic fluctuation time of a source provides an important indication of its compactness.

The majority of active nuclei emit most of their radiation from a region between one and several hundred Schwarzschild radii. A particularly remarkable active nucleus, called OX 169 has a luminosity (observed in the X-ray band) which can triple in 100 minutes, implying that the energy source is the size of Saturn's orbit. It is obvious that a special type of compact energy source is required to feed quasars.

[15] This technique uses groups of radio telescopes separated by thousands of kilometres (in different continents); interferometry then gives very high resolution.

Cosmic jets

At a distance of 16 million light years, Centaurus A is the closest radio galaxy. It is not a very powerful source, but it has two very fine beams of ionised gas which burst from each side of the centre of the galaxy and extend far beyond its optical frontiers to a distance of a million light years. These *cosmic jets* end in clouds, called *lobes*, which emit synchrotron radio signals.

In the visible band Centaurus A is a very beautiful object, resembling an elliptical galaxy hidden behind a layer of dust. Its nucleus contains a tiny variable radio source, whose size is less than a few light hours. Although its radio power is quite low, the quantity of energy which must have been injected into the lobes is equivalent to several million supernova explosions, which indicates that Centaurus A has been very active and has a central engine of at least 10 million $M_{\odot}$.

This double jet structure is not unique to Centaurus A; it is one of the remarkable characteristics of radio-emitting active nuclei. During the last few years, long-baseline interferometry techniques have enabled astronomers to resolve the jets into different sub-units encased within each other like Russian dolls. Microjets of several light years escape from the compact nucleus and are perfectly aligned with the giant jets which stretch out a million times further. This striking alignment of the gas structures over such long distances implies that the jets are expelled from a central engine which has 'remembered' their direction for millions of years. In other words, the cosmic jets are a giant version of those observed in stellar systems such as SS 433 (see page 254). These arguments reinforce the idea that there is a massive rotating compact body acting as an engine, the rotational axis controlling the direction in which the gas escapes.

Martin Rees has suggested that the most variable active nuclei are those in which the 'exhausts' by chance point directly at the Earth. The idea is to explain a disconcerting class of active nuclei called the Lacertids.[16] Their most striking property is that they

[16] The first galaxy of this type was discovered in the constellation Lacerta; originally it was classified as a variable star until it was identified as an extragalactic radio source in 1968.

exhibit luminosity fluctuations which are more rapid and more intense than other active nuclei. They fluctuate so quickly – every few hours – that their radiation appears to come from a region smaller than the size of a black hole of the same mass. There is another important difference: whereas the spectra of active nuclei are characterised by very intense emission lines,[17] the Lacertids have practically 'virgin' spectra. However, it is thought that the emission lines originate in a vast gas cloud surrounding the central source and illuminated by it, and that all active nuclei must possess these clouds.

Rees' model involving a jet pointing towards the Earth simultaneously explains the apparently too-rapid variations of the Lacertids and their lack of emission lines. In fact the theory of Special Relativity predicts that, if a jet of matter moves at a velocity close to that of light in the direction of the observer, its luminosity is amplified and the apparent fluctuation time shortened.[18] We can also understand that if the Lacertid jet is directed towards the Earth, the emission lines from the deepest layers are completely drowned out by the very intense radiation from the jet.

Continuity with change

The observations of 'normal' galaxies – in which the nucleus is less luminous than the rest of the galaxy – show them to have many characteristics in common with active nuclei. The most obvious illustration is the Milky Way, whose nucleus is a radio source associated with a high concentration of mass. It seems reasonable to suppose that the active galactic nuclei are not exotic monsters but are galaxies at a stage of their evolution giving favourable conditions for central activity.

The most important property of an active nucleus is the presence of a strong mass concentration. The observational methods mentioned above enable us to estimate the value of this mass for

[17] Whose redshift, for a quasar, is used to determine the distance.

[18] It also explains why certain jets seem to move at velocities greater than that of light.

any nearby galactic nucleus, provided it is not hidden by dust. The application of these methods to the study of nearby galactic nuclei over the past few years has produced and is continuing to produce a very surprising result: the presence of a central mass seems to be a common feature of nearly all galaxies, whether spiral or elliptical, giant or dwarf. Of the many examples, I shall discuss two particularly interesting ones.

Our Galaxy belongs to a group of about 20, dominated by the Andromeda nebula, which is visible to the naked eye just 2 million light years away. The Andromeda galaxy is a close relative of the Milky Way; it is a spiral galaxy, almost 1.5 times larger; it has the same chemical composition and small satellite galaxies. Since the plane of its disc is inclined with respect to the observer's direction, its central nucleus – non-active – is accessible to optical telescopes, and it is possible to measure the distribution of nearby stars. The most recent calculations have shown there is a dark central mass of 10 million $M_{\odot}$. This galaxy has a good-sized central engine, but it does not work!

One of Andromeda's satellite galaxies is Messier 32, a dwarf elliptical, 100 times less massive and totally inactive. It is simply a big swarm of old stars orbiting the centre. The absence of gas and dust makes it possible to study the nucleus with great precision: we can calculate that the stars are orbiting an invisible central mass of 5 million $M_{\odot}$. This dwarf has a heart as large as the Milky Way's.

We see from this that if giant black holes are present in the nuclei of most galaxies, their level of activity depends on the density of stars and gas – in other words, the 'fuel' – within a radius of a few light years. In this respect it is hardly surprising that Messier 32, despite its enormous engine, is totally inactive: the elliptical galaxy around it is completely devoid of gas and so small that it does not contain many stars. At the other extreme is Messier 87, a giant elliptical galaxy possibly containing a black hole of 5 billion $M_{\odot}$. There is a certain amount of activity in the nucleus, but much less than that of a quasar. To account for its luminosity it needs 0.01% of a solar mass of gas to be sucked towards the black hole every year. This modest amount of matter can easily be supplied by the millions of nearby stars, which lose gas as a normal part of their nuclear evolution. Messier 87 could be perhaps an extinct quasar.

About a billion years ago, the quasar would have been operating at full throttle; at 50 million light years from the Earth, it would have been visible to the naked eye at night, with a brightness similar to that of Mercury.

Alternative engines

The detection of high concentrations of matter in the nuclei of active galaxies is not definite proof of the presence of giant black holes. *In principle*, two other objects could act as compact and efficient engines: an ultradense star cluster, or a single supermassive star. Do these competing models withstand closer analysis? We shall see that they do not.

The idea of a star cluster is based on the exceptional rate of supernova explosions that would occur within the cluster. Supernovae, the natural result of the thermonuclear evolution of massive stars, are statistically quite rare in a galaxy – a few per century. However, in a very dense cluster the frequency of explosions would increase because of *star collisions*. In fact the collision of stars would usually lead to the merging of the two partners to form a massive single star, which would then evolve more quickly towards the supernova state. Calculations suggest that in a very dense cluster containing a billion stars collisions are so frequent and the number of massive stars formed is potentially so high that there could be 10 supernova explosions per year.

There are three major problems with the star cluster model. In the first place it does not explain the large luminosity variations of quasars and Lacertids. Each supernova explosion is a burning match besides the blazing flames of a quasar. To reproduce the quasar fluctuation, 1000 supernovae would have to explode simultaneously. Secondly, a star cluster is incapable of producing large stabilised cosmic jets because it does not define a particular direction in which to expel matter. But the most serious objection is the extreme *instability* of dense star clusters; a group of a billion stars confined within a light year – the observational constraint – would survive only a million years before collapsing into a black hole. It would need an extraordinary coincidence for active galactic nuclei to be observed systematically in such a very brief and special phase of their evolution. Once more *Occam's Razor* intervenes, ruling out a dense star cluster as the engine in active galactic nuclei.

The idea of a supermassive star fares no better. The theory of stellar structure explains why no observed star has a mass greater than 100 $M_{\odot}$. However, astrophysicists still speculate regularly on the existence of supermassive stars of 100 000 to 100 million $M_{\odot}$. The main feature of a supermassive star is that it should be extremely bright. But this is also its problem: a supermassive star is just a huge photon sphere, and photon spheres are not stable systems. Even if supermassive stars were formed by some mechanism not yet understood, they should either explode or collapse.

Since this does not work, several varieties of supermassive star have been invented in the hope of producing large but stable masses. 'Spinars' are rapidly rotating supermassive stars whose equilibrium is maintained by centrifugal forces. 'Magnetoids' on the other hand are stabilised by colossal internal magnetic pressure. These theoretical stars, resembling giant pulsars, have the merit of possessing a privileged direction for ejecting matter – the rotational or magnetic field axis. However, the theory of General Relativity shows that they are fundamentally unstable, mainly due to the dissipation of energy by gravitational waves. Furthermore, a giant pulsar would produce periodic luminosity variations, which have never been observed in any galactic nucleus.

In conclusion, the massive accreting black hole is the only candidate which fulfils all the theoretical and observational requirements for explaining the activity of galactic nuclei. Its formation is predicted by General Relativity, and is, with little doubt, the end product of the gravitational collapse of all massive objects. A black hole is stable and provides an ideal environment for converting gravitational potential energy into radiation, by accreting matter. Finally, the black hole is not only able to liberate energy from the matter which falls into it, but is itself a large reservoir of rotational energy (see Chapter 11). Since its rotational axis provides a preferred direction for the ejection of matter, a rotating black hole can function as a generator of gas jets, like the star SS 433 but on a much greater scale (see Figure 66).

Let's eat!

For an average efficiency of 10% the active nucleus of a galaxy, from the most modest to the brightest, needs to consume between 0.01 and 100 solar masses of gas per year. What could supply this amount of matter?

Year-in, year-out, in a spiral galaxy like the Milky Way, the stars expel 1 solar mass of gas. It is difficult to understand how this gas, spread throughout the 100 000 light years of the disc, can be channelled into the tiny central nucleus 1 light year in diameter. On the other hand, some elliptical galaxies, although stripped of interstellar gas, also manifest signs of activity – mainly the emission from radio jets.

There must therefore exist a more radical mechanism which produces large quantities of gas in the nucleus itself. Since the gas is contained in the stars, we can conclude that in order to feed itself *a black hole must break up stars.*

Giant black holes are perfectly capable of swallowing whole stars. To a giant black hole the Sun would be like a grain of sand next to a football. However, this way of eating does not release energy; it would all be lost within the black hole, which would slightly increase in mass. In order for energy to be released the star must be broken up outside the black hole, so that some of its fragments can become part of the accretion disc.

Cometary stars

In many respects a giant black hole immersed in a star cluster resembles the Sun surrounded by its retinue of comets. The stars in the cluster act as a sort of reservoir, far from the black hole and scarcely influenced by the central gravitational field. However, some of the stars in this reservoir skim so close that they are accelerated and thus deviated from their trajectory. Sometimes one of them plunges directly towards the black hole. From this moment its destiny is entirely controlled by the gravitational well which attracts it and the radiation field which illuminates it. Like a comet approaching the Sun, the star receives an intense flux of radiation, coming not from the black hole but from the hot regions of the

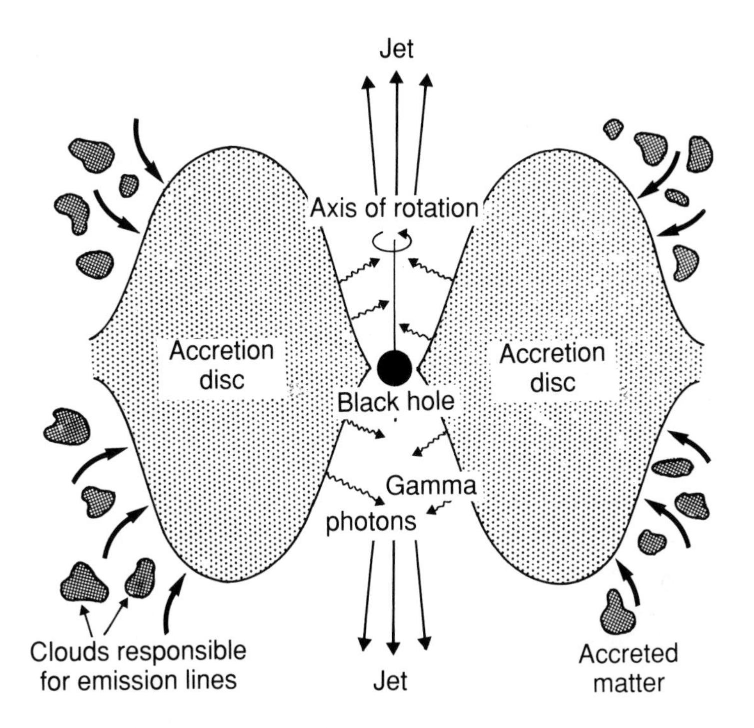

Figure 66. Schematic representation of a thick accretion disc around a giant black hole with the formation of jets. Gamma-ray photons emitted by the internal face of the disc create electron–positron pairs in the vacuum, which multiply by cascading (see page 102) and accelerate the jets to a velocity approaching that of light.

accretion disc which surround it. The star begins to evaporate, shedding its outer layers until the thermonuclear core is left naked. If the star does not pass too close to the black hole it may come out of the gravitational well on a parabolic orbit without too much damage, and after a few years return to its original position; or the 'cometary' star might lose so much orbital energy that it remains bound to the black hole on an elliptical orbit which draws it closer and closer to the central source, losing some gas at each 'perihelion' passage.

However, the evaporation of cometary stars would provide only a modest contribution to the black hole's food supply. To shine actively the black hole must consume the equivalent of an entire star in gas. There are two events capable of disrupting the entire mass of a star in a suitable form. One is the *collision* between two cometary stars near a black hole, the other is the disruption of an individual star by the *tidal forces* of a black hole.

Star–star collisions

The chances of 2 comets colliding in the Solar System is extremely slight; not so, near a black hole. The theory of interstellar collisions shows that when two solar-type stars collide at a low velocity, less than 500 km/s, the shock is 'soft' and they remain stuck together as a single large star. If their velocity is greater than 500 km/s, the shock is harder and stellar fragments are scattered widely. In the galactic disc or even stellar clusters, stars rarely have velocities greater than 200 km/s. However, giant black holes produce such deep gravitational wells that nearby stars are accelerated to velocities of several thousand kilometres per second. We can calculate that in a radius of 10 light years around a black hole of a billion $M_{\odot}$, collisions between cometary stars are destructive and occur at a rate of 10 per year. Their fragments remain in orbit around the black hole in the form of gas clouds, and fill the black hole's 'larder'.

However, it is likely that interstellar collisions are able to supply only those quasars associated with very large black holes; for the less active nuclei associated with smaller black holes, the frequency of interstellar collision is so low that they probably play no role at all.

Black tide

The most impressive phenomenon occurring around a giant black hole is perhaps the *break-up of a star by tidal forces*. As a star moves around a black hole the forces of gravity act more strongly on the side of the star nearer the black hole than on the other side. The difference between the two forces is the tidal force exerted by the black hole (see page 37). If the star moves in an approximately circular orbit the tidal force remains small and the star is able to adjust its internal configuration to the external forces, adopting an elongated shape, oriented towards the black hole. However, if the star is moving in an eccentric orbit within the black hole's gravitational field, the tidal forces increase rapidly as the distance from the black hole decreases.[19] Eventually there comes a point where these forces are as large as the forces binding the star together. The star no longer has time to adjust its internal configuration, begins to deform catastrophically and is inevitably disrupted.

These spectacular events occur only when a star gets within a certain critical radius from the black hole, called the *Roche limit*, after the French mathematician who studied the problem of tidal forces in 1847 in the context of planets and their satellites.[20]

The Roche limit depends mainly on the mass of the black hole. If it is too large – more than 100 million $M_\odot$ – its own radius, which is proportional to its mass, is greater than the Roche radius. In this case the star can only be broken up by the tidal forces once it is inside the black hole. All the debris would then be sucked into the black hole and astronomers would not observe anything. For smaller black hole masses the star can be disrupted by tidal forces outside the hole. This is why most astrophysicists today believe that Seyfert galaxies and almost inactive galactic cores have a central black hole of between 1 million and 100 million $M_\odot$, feeding on the remains of stars broken up by tidal forces, while quasars and the bright nuclei harbour a more massive black hole, for which interstellar collisions supply the fuel.

Flambéed pancakes

The description of the deformation and break-up of a star by tidal forces has long been based on Roche's description of a liquid or solid satellite in a circular orbit around a planet. He

[19] The tidal forces become infinite at the centre of the black hole, (see page 130).

[20] Edouard Roche also gave his name to the 'Roche lobes' which characterise the regions of gravitational influence of a binary star, see Figure 59. It is interesting to note that when a star crosses the Roche limit it is broken up as easily as if it had been involved in a collision with another star at a velocity in excess of 500 km/s. When a star penetrates the Roche limit it is as if it collides with itself!

showed that a body which is subject to the tidal forces of a massive neighbour tends to elongate in the direction of its companion and to compress in the perpendicular direction. This is why ocean levels are higher not only at the point nearest the Moon, where the attraction is strongest, but also at the diametrically opposite point (Figure 67). If the tidal forces become quite large – as in the case of some very tightly bound binary systems – the bodies deform lengthways to become 'cigar' shaped. The Roche limit marks the point beyond which the deformations are so great that the body becomes unstable and starts to break up.

While this is true for a planet–moon situation, it is not necessarily the case for a black hole–star system, involving quite different types of object. At the Meudon Observatory, Brandon Carter and myself decided several years ago to re-examine the question; we discovered unexpected phenomena which challenged some of the accepted ideas about the disruption of celestial bodies.

A black hole–star system differs from a planet–satellite system in two main ways. In the first place the orbit of a cometary star is not circular but is stretched in the direction of the black hole. In fact for a star to reach the region where the tidal forces are destructive it has to be travelling on a very eccentric orbit. If the Galactic Centre contains a black hole of 3 million $M_{\odot}$ whose radius is 10 million kilometres, all stars like the Sun approaching within 200 million kilometres would be destroyed; for this is the Roche limit for the Galactic Centre. The question we asked ourselves is as follows: what would happen to a star which instead of skimming the Roche limit went *deep* inside it, without being swallowed by the black hole? After all there is a lot of space between the 10 million kilometres of the black hole's surface and the 200 million kilometres of the Roche limit. The tidal forces vary as the inverse of the cube of the distance from the black hole. This means that at a distance 10 times smaller than the Roche limit the forces are 1000 times greater than at the Roche limit, where they are already strong enough to destroy stars. Thus a star reaching this distance is likely to suffer a much more violent fate than those barely brushing the Roche limit.

The second important characteristic of the black hole–star system is the nature of the object subjected to the tidal forces: an ordinary star like the Sun, unlike a moon or planet, is made of gas

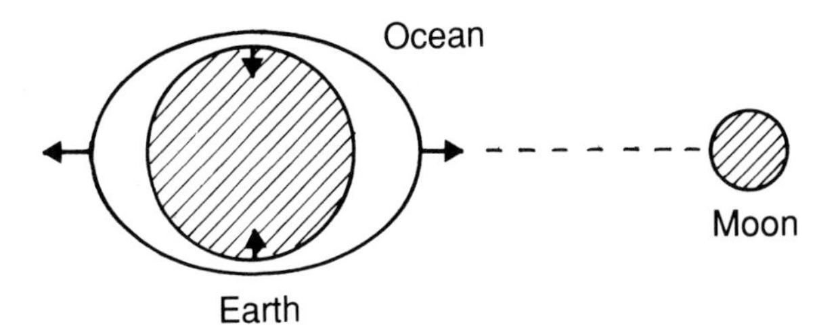

Figure 67. Tides on the Earth.

The Moon's gravitational forces deform the ocean surfaces, producing a bulge in the Earth–Moon direction and a narrowing in the perpendicular direction. In reality the tides are not exactly aligned with the Moon, because of the rapid rotation of the Earth.

and not of rock; it is therefore more easily *compressed* by the tidal forces. This is precisely what happens when a star plunges deep within the Roche limit of a giant black hole. Although initially it tends to assume a cigar shape, the tidal forces act upon it like a giant mangle and flatten it out within the orbital plane (Figure 68).

Compression means heating. These two factors depend very sensitively on the depth of penetration within the Roche limit. If the star skims just within the limit the forces are not strong enough to compress it; the star behaves as if it were a giant ball of water, elongating into a cigar and swelling up, and finally breaking up after it emerges from the Roche limit. If on the contrary the star reaches a distance 10 times less than the Roche radius it will be so compressed by the tidal forces that its density will increase by 1000 times and its temperature by 100 times in a tenth of a second. Of course the star ultimately breaks up and its gas dissipates, but before this it resembles a *giant flambéed ultra-hot and dense pancake*.

The black hole detonator

The most spectacular consequence of the crushing of a star is the release of a thermonuclear explosion in the 'stellar pancake'. The rate of nuclear reactions governing the flow of energy depends very much on temperature. For a star in hydrostatic equilibrium,

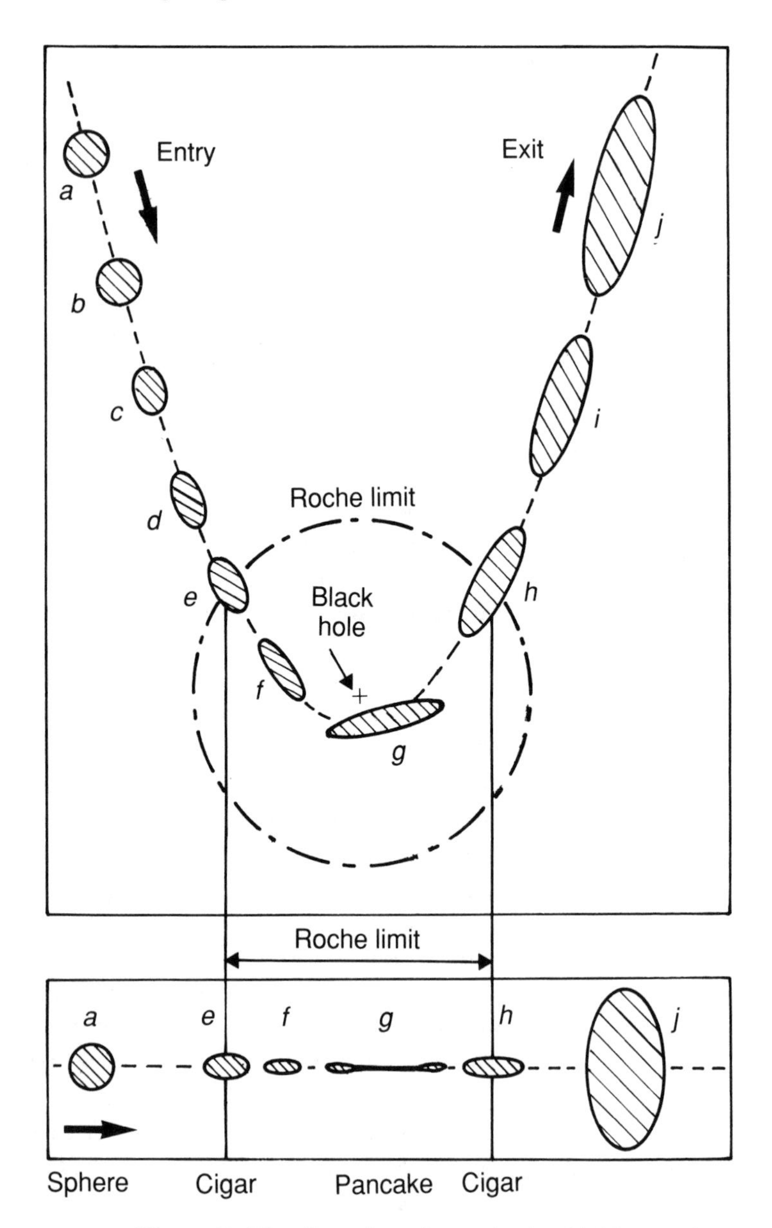

Figure 68. The disruption of a star by the tidal forces of a black hole.

The diagram illustrates the progressive deformation of a star plunging deep inside the Roche limit of a giant black hole (the

like the Sun, the central density is 100 g/cm^3 and the temperature is 15 million K. In these 'normal' conditions, the dominant nuclear reaction is hydrogen fusion, but this occurs at an extremely slow rate (see Chapter 4).

If a star chances to plunge within the Roche limit of a giant black hole, its central temperature increases to a billion degrees in a tenth of a second. As in the phases leading up to a supernova explosion, the thermonuclear chain reactions are considerably accelerated. During this brief period of heating the hydrogen in the star is unable to fuse, but the heavier elements like helium, nitrogen and oxygen, which previously were inert, are instantaneously transformed into heavy elements, releasing energy. A thermonuclear explosion takes place within the stellar pancake, resulting in a kind of 'accidental supernova'.

The consequences of such an explosion are far reaching. Part of the stellar debris is blown away from the black hole, beyond its range, as a hot cloud capable of carrying away any other clouds it might collide with. The rest of the remains fall rapidly towards the black hole, producing a short burst of radiation. Like supernovae, the stellar pancakes are also crucibles in which heavy elements are produced and then scattered throughout the galaxy. However, calculations show that the proportions of the elements produced by a stellar pancake are slightly different from those produced during a supernova. It may therefore be possible in the near future to detect these elements in the spectrum of an orbiting cloud near the

Caption for Fig. 68 (*cont.*)

size of the star has been considerably exaggerated for clarity). The upper diagram represents the deformation of the star in its orbital plane (seen from above), the lower one shows the deformation in the perpendicular plane (seen from the side). From *a* to *d* the tidal forces are weak and the star remains practically spherical. At *e* the star penetrates the Roche limit. It becomes cigar-shaped. From *e* to *g* the 'mangle' effect of the tidal forces becomes important and the star is flattened in its orbital plane to the shape of a 'pancake'. The star rebounds, and as it leaves the Roche limit, it starts to expand *h*, becoming more cigar-shaped again. A little further along its orbit *j* the star finally breaks up into gas fragments.

centre of an active galaxy, and this would give direct evidence of the explosive rupture of stars by a giant black hole.

Gravitation is the cause of thermonuclear explosions in both supernovae and stellar pancakes. In a supernova the gravitational field of the star itself causes it to become unstable and detonates the explosion through collapse of the core. In a stellar pancake the black hole's gravitational field compresses the star from the outside and causes it to explode.

This type of event, where a star explodes in the extreme gravitational forces caused by a black hole, is rare. We can calculate that of the limited number of stars approaching within the Roche limit for a black hole – about 1 per year in active nuclei, and 1 every 1000 years in the Galactic Centre – only a tenth of these will penetrate far enough to explode. But tides are not the only means of tossing these stellar pancakes. Head-on collisions between stars travelling at very high velocities, which happen quite frequently – about 10 per year – in the vicinity of supermassive black holes of a billion $M_{\odot}$, also form transient star pancakes. The phenomenon could therefore play as important a role in almost inactive galaxies, where the moderately massive central black hole would produce pancakes by tidal effects, as in quasars, with larger black holes, where they are produced by the collision of stars.

Cannibalism

Knowing that the internal combustion engine can produce energy does not mean that we understand the detailed working of a car. Although the model of a giant black hole as the engine at the heart of an active galactic nucleus is plausible, we should recognise that the details of the activity are poorly understood, and quasars remain one of the most mysterious phenomena in the Universe.

Observation of distributions of quasars, rather than of single quasars, has provided us with much information on their formation, extinction and role in the life of the galaxies which shelter them. We should first ask whether all galaxies pass through the quasar stage at one time or another in their evolution, and what stage produces this spectacular phase. Ideas about this are somewhat contradictory. Quasars are observed only in very distant galaxies, i.e. a long time ago, which suggests that they belong to a primitive phase of galactic evolution. On the other hand, if the giant black holes needed to feed quasars are developed from a stellar seed, this suggests that the quasar stage is an advanced one in the evolution of galaxies, and that most galaxies have not yet passed this stage.

If all galaxies are going to pass through an active phase at one time or another, the fact that relatively few quasars are observed indicates that this phase is short, lasting a few tens of thousands of years only. But the observation of extended radio jets suggests that the lifetime of the central engine cannot be as brief as this, because the jets would not have been able to align themselves over such great distances. But the quasar phase could hardly last much longer because of the insoluble problem of fuel supply. We conclude that the active quasar phase must last about 100 million years and occurs only in a minority of galaxies where exceptionally favourable conditions hold temporarily. In this scenario the quasar phase would be activated as soon as the black hole reached a sufficiently high mass and there was a substantial supply of gas and stellar 'fuel' available; it would cease once this fuel supply had fallen below a certain level. In this way there would be more dead quasars than live ones.[21]

A dead quasar could be revived, given a new supply of food. Once the central star cluster has been used up, it needs to extract matter from elsewhere. Now galaxies encounter each other fairly often, especially those in clusters which contain many hundreds or many thousands of members. Recent observations show that many quasars are associated with *colliding galaxies*, suggesting that like stellar X-ray sources, activated by accretion from a companion star, the activity within galaxies can be stimulated by the transfer of matter between partners.

It has also been noticed that among nearby galaxies, those which form part of a multiple system are slightly more active than galaxies that are isolated. This phenomenon is particularly noticeable at the centre of clusters very rich in galaxies, which contain

[21] Quasars are situated billions of light years away, and their lifetime is scarcely 100 million years; all the quasars now observed became extinct a long time ago.

'supergiant' elliptical galaxies 100 times larger than 'normal' galaxies. These galaxies are very active in the radio region and are surrounded by a host of satellite galaxies which are falling into them. These supergiant galaxies therefore grow by a sort of *cannibalism*, swallowing nearby smaller galaxies. It is reasonable to imagine that a number of the captured galaxies have a large black hole at their centre; so the cannibal galaxies probably have a multiple nucleus composed of several massive black holes perturbing the distribution of the surrounding matter and increasing the accretion rate. In fact several centres of activity have been observed in the giant radio galaxies. However, the ultimate destiny of a group of black holes is to merge into a single giant hole larger than the sum of its components; so one day in the distant future, the black holes will run out of food and active galaxies will be extinguished.

In the meantime, if black holes are the giant engines of quasars we have arrived at a strange paradox: a black hole, perfectly invisible to the naked eye, becomes the brightest source in the Universe if it is suitably clothed in the gas of stars.

18

Gravitational light

'I should like to put to Herr Einstein a question, namely, how quickly the action of gravitation is propagated in your theory.'
Max Born (1913)

In Newton's theory gravitation is a force which acts instantaneously between massive bodies. This idea, as we have already seen was unacceptable to many physicists, including Newton himself. A century later Laplace modified this theory so that gravitational interaction propagated at a finite speed. This idea was soon abandoned because it raised a question no-one could answer: if a massive body is subjected to a violent perturbation, its gravitational field must adjust itself over a short period of time to match the new configuration of the body; how does this readjustment propagate?

Einstein's theory of General Relativity gives a consistent picture of the propagation of gravitation. Einstein asked himself if an accelerating mass could radiate gravitational waves, just as an accelerating electric charge radiates electromagnetic waves. In 1918 he discovered solutions of the gravitational field equations representing *waves of space-time curvature travelling at the velocity of light*. He had just invented 'gravitational light'.

The analogy between gravitational and electromagnetic waves is useful for understanding the concept behind the phenomenon, but not much more. The structure of a gravitational wave and its effect on matter are much more complex than those of the electromagnetic wave. The first important difference is that gravitation is only attractive; the mass, or the 'gravitational charge', always has the same sign. The outcome of this is that an elementary gravitational 'oscillator', consisting of two masses vibrating on either end of a spring, does not radiate the same type of waves as two electric charges of opposite signs.[1]

Another complication is that the *graviton*, the hypothetical carrier particle of a gravitational wave, carries a 'gravitational charge' associated with its energy, whereas the photon, the carrier particle of electromagnetic interaction, does not have an electric charge. Consequently the gravitational wave produced by an accelerating mass is itself a source of gravitation: gravitation gravitates. In technical terms we say that it is 'nonlinear'. This nonlinearity introduces considerable difficulties into the solution of what appears to be even the simplest problem, such as the calculation of the gravitational field generated by two moving bodies. Unlike the electromagnetic field, if there are two masses generating their own gravitational fields, then the combined field is not the sum of the two individual fields; we have to take into account the *gravitation of the interaction* of the two masses, which varies constantly as they move. This is why the 'two body problem', describing for example the gravitational field of a binary star system, for which the Newtonian solution is easy to calculate, cannot be solved rigorously in General Relativity.

However, in sufficiently weak gravitational fields the 'non-linearity' can be ignored and the problem simplified. This is true if we wish to detect gravitational radiation from distant sources. However, these simplified equations cannot be applied in the vicinity of a supernova or two colliding black holes.

The third fundamental difference between gravitational and electromagnetic waves is their relative intensity. Two protons placed a centimetre apart have masses and electric charges, and are subject both to a gravitational interaction and an electromagnetic interaction. But the gravitational force which attracts them is 10^{37} times smaller than the electrostatic force which repels them.[2] This is the major obstacle hindering the detection of gravitational waves; although Hertz managed to generate and receive electromagnetic

1 In electromagnetism the radiation is *dipolar*, whereas in the gravitational case it is *quadrupolar*. Individually, each of the masses of a gravitational oscillator acts as a dipole which is incapable of generating gravitational waves.

2 In an atomic nucleus, the nuclear interaction which links two protons is 100 times stronger than the electromagnetic interaction.

waves in the laboratory only a decade after Maxwell had predicted their existence, 70 years have elapsed since Einstein predicted the existence of gravitational waves and no-one has yet detected them.

There are several more examples which illustrate the extreme weakness of gravitational waves under normal conditions. Let us reconsider our elementary gravitational oscillator consisting of two 1 kilogram masses oscillating through 1 centimetre 100 times per second at the ends of a 10 centimetre spring. Assuming that all the gravitational power released by this system is converted into electric power, the number of such oscillators required to light a 50 watt bulb would be more than the number of particles contained in this planet!

Another way of constructing a gravitational oscillator would be to rotate a horizontal bar about a vertical axis passing through its centre. In the rotational plane the bar appears to contract and expand as it shows first one of the ends and then one of the sides to the observer. This action produces gravitational waves. If we take a steel bar 20 metres long and weighing 500 tonnes, and rotate it at its maximum speed within its stress limit, which is 5 turns per second, the gravitational energy released is again ridiculously small: 10^{-29} watts.

We would do better to leave the laboratory and look for natural sources in the Solar System, but the situation is not much more encouraging. It would require 50 billion meteorites 1 kilometre in diameter crashing into the Earth at 10 km/s to light a modest light bulb. There would be no-one left alive to see the result.

It is useless to look for sources in ordinary astronomical bodies. To produce non-negligible gravitational waves, a star has to be travelling at nearly the velocity of light and must be compact, that is close to its Schwarzschild radius. The Earth, which revolves around the Sun with a velocity of 30 km/s and whose radius is a billion times greater than its Schwarzschild radius, produces a gravitational power of just 0.001 watt.

Throughout this book we have talked about 'relativistic' stars, which at least temporarily can reproduce the conditions favourable to the emission of gravitational light. Only the most violently perturbed astronomical sites are good generators of gravitational waves. Since these stars are very far away (if such an astronomical cataclysm occurred near the Earth, it would destroy all life), only a tiny fraction of their gravitational energy reaches the Earth.

Compact star systems are good generators of gravitational waves. A closely coupled pair of neutron stars radiates enough gravitational energy for its effects to be detected indirectly: the loss of orbital energy is indicated by a decrease in the period of revolution. The binary pulsar PSR 1913 + 16 is a perfect illustration of this phenomenon. At present it is probably the only observational demonstration of gravitational waves (see page 107).

For individual stars the catastrophic events which herald the end of their thermonuclear life could be powerful sources of gravitational radiation. A supernova which results in the formation of a neutron star may be extremely efficient. A star could emit more gravitational energy in the last few seconds of its collapse than all the electromagnetic energy released during the millions of years of its thermonuclear life. However, unlike binary stars which emit periodic gravitational waves and could be labelled 'gravitational pulsars', a supernova is an 'impulsive' source which produces a single brief burst of gravitational radiation.

A discussion of gravitation always leads back to the black hole. It is the relativistic star *par excellence* and the most prolific source of gravitational radiation. The perfectly spherical collapse of a star into a black hole does not produce any waves (see Chapter 11), but real stars rotate and there are always asymmetrical motions which allow gravitational light to escape. The first 'cry' of a baby black hole is a flash of gravitational light carrying an amount of energy comparable to its rest mass energy. The gravitational luminosity from the collision of two 10 $M_\odot$ black holes is 100 million times greater than the electromagnetic luminosity of the most powerful quasar! If such an event happened at the centre of our Galaxy, 10 000 light years away, a detectable flux would reach the Earth.

Thus a new astronomy is emerging, the study of gravitational light. This will be an astronomy of incomparable transparency, because, unlike electromagnetic radiation, gravitational radiation is not absorbed by matter, and radiation from distant sources can reach the Earth without losing any of the information it carries. In addition, the strongest sources are those about which electromagnetic observations reveal little and that only in an indirect way:

pairs of neutron stars, the cores of supernovae and black holes. Thus gravitational astronomy will open the window on an even more mysterious Universe, not only revealing unknown properties about compact stars and ultradense matter, but also telling us about the beginnings of the Universe 15 billion years ago. The primordial Universe, continuously agitated by density fluctuations, and the Big Bang itself, were powerful sources of gravitational radiation; and even if no electromagnetic waves emerged during the first million years following the Big Bang, gravitational radiation would have passed without hindrance through the densest regions of the primordial Universe. Gravitational light will perhaps provide the only definitive proof of the existence of black holes and of the birth of the Universe.

Let us return to the Earth. Telescopes are used to capture light; how do we build a gravitational telescope? The principle is simple. Just as electromagnetic waves cause the receiving antenna to vibrate, gravitational waves cause the matter they encounter to vibrate in a certain way; the 'wrinkles of curvature' cause the elastic fabric of space-time to undulate slightly, lengthening or shortening distances as they pass by. If, for example, the detector was a block of solid matter, its different parts would move in different directions as the gravitational wave passed through.[3]

The relative displacement of two masses with respect to their separation defines the *amplitude* of such a wave, which is a direct measurement of its power. A collision between two stellar black holes at the centre of the Galaxy would cause a displacement of 10^{-12} millimetres (a thousand billionth) between the two extremities of a bar-shaped detector 1 metre long. The construction of a gravitational wave detector is therefore a technological challenge to scientists.

In the 1960s, Joseph Weber at the University of Maryland constructed large cylinders of aluminium which were designed to respond to gravitational waves from the Galactic Centre by oscillating. He thought that he had obtained positive results; however, as other similar experiments throughout the world

[3] It should be noted that as gravitational waves are always passing through any object, none of them, no matter how rigid, is strictly undeformable.

demonstrated, he had incorrectly interpreted the experimental errors. A supernova explosion in the Galactic Centre would produce a wave with an amplitude of 10^{-18}, whereas the best Weber bars could only detect an amplitude which was 10 000 times greater. In addition, the detection of supernovae in the Galactic Centre would cause a problem: in the Galaxy there is one supernova every 10 years, and the gravitational pulse of such an explosion would last only a fraction of a second.

The most favourable site for the detection of gravitational waves is in the galaxy cluster in Virgo, where supernova explosions and the decay of binary pulsars from the several thousand galaxies grouped together in a small angular region of the sky occur at a rate of about one per week. But the Virgo cluster is not 10 000 light years away like the Galactic Centre, but 50 million. This means that to detect gravitational light from one of these supernovae, the gravitational telescopes would have to be a million times more sensitive than one capable of detecting a similar event in the Milky Way. It is worth noting that the explosion of the supernova in the Large Magellanic Cloud in February 1987 (see Chapter 6) was 'only' 170 000 light years away, and could have emitted a burst of gravitational waves strong enough to have been detected by the two or three detectors in operation . . . if they had been switched on. But, on that day they were being serviced!

Despite these discouraging technological difficulties, the challenge of detecting gravitational light may be met by the end of the century. Many technical advances have been made since Weber's time. At present there are eight research groups from several countries working on second-generation bars, which are not only more sensitive but also more costly since they are constructed from rare materials such as niobium or sapphire cooled to a few degrees above absolute zero.

Another more promising path has just been opened. The principle consists of measuring oscillations in the separation of two massive mirrors placed at the ends of long arms whose separation is controlled by a system of light interferometry. This is a modification of the Michelson–Morley experiment (see Chapter 2), no longer used for detecting the absolute motion of the ether, but for measuring the trembling of space-time. The greater the distance between the two mirrors, the greater the chance that the effects of the gravitational signal will be detectable amid the 'background noise' inside the system (caused by seismic waves, sound waves and so on). The construction of very high quality mirrors capable of several hundred successive light reflections should make it possible to obtain distances equivalent to 150 kilometres with a real separation of 3 kilometres.

No interferometric antenna has been built yet, but various preparatory experiments are currently under way: a US project, an Anglo-German project, and the Franco-Italian project VIRGO (named after the Virgo cluster of galaxies, its main target). The amount involved is less than the cost of a single aircraft or a satellite launch . . . or half an hour of war in the Persian Gulf. However, gravitational astronomy, lacking observational evidence, does not receive money as a matter of course. The community of 'relativistic' astrophysicists is still waiting anxiously to see if they will get the funding to help them open another marvellous window on the Universe. The recent history of astronomy has shown that each time we examine the heavens through eyes other than our own or a camera's (radio telescopes and gamma-ray and X-ray detectors), new marvels are found, forcing us revise our ideas and improve our understanding of the Universe.

The gravitational window to the Universe will be opened sooner or later. When the first signals are detected, the information about the motion and nature of these sources will still be drowned in the background noise. However, sustained by the certainty that gravitational astronomy is that of the coming centuries, we could perhaps attempt to launch giant gravitational interferometers into space, insulating them from earthly and human disturbances.

19

The black hole Universe

'Eternity is very long, especially near the end.'

Woody Allen

In this final chapter it is time to place black holes in cosmic perspective. We have looked for light from microscopic primordial black holes smaller than an atom, we have witnessed the birth of stellar black holes with a radius of 10 kilometres, and we have rubbed shoulders with giant black holes the size of the Solar System. There is just one question left to ask: *what is the biggest black hole possible*? The answer is one of the most fantastic hypotheses in modern science: *the Universe itself*!

To understand why such a concept is not madness, we have to recall some elements of cosmology. Beyond the myths and wild imaginings that human beings have thrown up in constructing a comprehensible and reassuring Universe, the modern cosmologist has three observational facts which, given careful physical interpretation, allow him to deduce the past history of the Universe. The motion of the galaxies, the relative abundance of light elements[1] and the detection of a uniform cosmic radiation all indicate that the Universe has been expanding for the past 15 billion years, ever since the very condensed and hot phase of the Big Bang.[2]

Observation has provided us with an insight into the history of the Universe, but only theory will enable us to guess at its future. Since gravitation dominates the organisation of physical structures on the large scale, Einstein's theory of General Relativity offers

[1] Hydrogen, deuterium and helium, which are not formed in stars.

[2] There are several excellent popular works which deal with this question.

plausible cosmological models consistent with the conditions which prevailed in the past. As for the future, there are two possible solutions: an expanding–contracting Universe, finite in time and space, and an indefinitely expanding Universe.[3]

The average density of matter in the Universe acts as the force deciding between these two destinies. If it is less than the critical value 10^{-29} g/cm^3 – equivalent to 6 hydrogen atoms per cubic metre of space – the universal gravitational field is not strong enough to keep matter bound, and the Universe will continue expanding indefinitely. Alternatively, if the average density is greater than the critical value, the gravitation will act on the expanding Universe and it will eventually recontract, so that in 100 billion years (10^{11} years) it will collapse into a sort of inverse Big Bang, the Big Crunch.

Whatever the final destiny of the Universe,[4] black holes will have a major role to play. Freeman Dyson, from the Institute for Advanced Studies at Princeton, and Jamal Islam, from London University, have studied the long-term evolution of a continuously expanding Universe.[5] Long duration physical processes which have not yet begun, though the Universe has been in existence for 15 billion years, will sooner or later appear on the scene. In about 10^{27} years all the extinguished stars will have gathered in the centre of the galaxies and formed massive galactic black holes of 10^{11} $M_{\odot}$. The motion of galaxies in the clusters will then disperse their orbital energy by gravitational radiation, and in about 10^{31} years the galaxies will fall into the centres of the clusters and fuse together into supergalactic black holes of 10^{15} $M_{\odot}$. On a much greater time scale the inverse process of 'quantum' disintegration of the black holes will take place. The stellar black holes will evaporate in 10^{67} years, the galactic black holes in 10^{97} years and the supergalactic

[3] Contrary to popular belief – and this includes some cosmologists – the fact that the Universe is infinite in future time does not mean that it is infinite in space; see my article on this subject 'Géométries de la variété univers', in 'Les confins de l'Univers', *La Nouvelle Encyclopédie des sciences et techniques*, Fayard-Fondation Diderot, 1987.

[4] Actual measurements of the density of the Universe produce a value which is slightly less than the critical density, but it is impossible to conclude in favour of the 'open' Universe because not all the matter has been found.

[5] These ideas are developed in J. Islam's book: *The Ultimate Fate of the Universe*, Cambridge University Press, 1983.

black holes in 10^{106} years. As the ultimate reservoirs of energy and entropy, black holes will become analogous to white holes releasing their matter into the external Universe.[6]

Dyson finally asked himself if, despite the unfavourable context of a Universe inexorably diluting and cooling itself, advanced civilisations might be able to maintain themselves indefinitely by extracting energy from black holes. This scenario is reminiscent of some classic science fiction stories, and is in conflict with one of the predictions of modern particle physics, according to which the proton is not eternal but disintegrates after about 10^{32} years.[7] The dissolution of physical structures and living systems would have occurred long before black holes started to release their energy.

Let us now examine the consequences of an expanding–contracting Universe, restricted in time and finite in space. The minimum density for the Universe to be a closed system is that of a 10^{23} $M_{\odot}$ black hole with a radius of 40 billion light years.[8] In the observable Universe, the largest distance travelled by light does not exceed 15 billion light years. This means that the observable Universe is inside its Schwarzschild radius! Can we conclude therefore that we are living within a gigantic black hole?

Considering the question more deeply, we see that a number of theoretical arguments support the hypothesis of a black hole Universe. A last mental effort and the reader may think back to Figure 47, the map of space-time inside and outside a collapsing spherical star. The exterior is a piece of Schwarzschild geometry, while the interior has a geometry depending on the equation of state of stellar matter. General Relativity shows that if the star is similar to a spherical 'cloud' of zero pressure and uniform density – i.e. similar to the galactic gas filling the Universe – then the interior geometry of the cloud (shown by the hatched area on the diagram) is strictly identical to that of a closed Universe, and the internal and external geometry connect perfectly at the cloud's surface.

[6] But as the 'graduated' radiation of a black body, see Chapter 15.

[7] Current experiments have not yet confirmed this prediction of the lifetime of a proton.

[8] It should be remembered that the average density of a black hole decreases as the radius increases.

On the other hand, the closed expanding–contracting Universe possesses an *event horizon*, that is, a boundary in space-time beyond which there are events which are always inaccessible because their light signals never reach us. This cosmological horizon[9] is associated with a future singularity (the Big Crunch) and *seen from inside* is the same as the event horizon which defines the frontier of the black hole *from the outside.*[10]

We might therefore imagine that if the Universe is closed, there should be an outside world within which our Universe is a region hidden inside a black hole. Clearly if this – still confused – hypothesis seems justified, a whole new field of cosmology will open up.

For example, scientists may wonder how our Universe became a black hole in the first place. Is it a primordial black hole in an exterior universe, or was it formed by the gravitational collapse of a $10^{23}\ M_\odot$ 'superstar'? In this case the exterior cosmos would not be empty and entire galaxies – perhaps formed from unknown matter – could fall into our Universe.

The most appealing consequence of the Universe as a black hole would be the completely unexpected behaviour of matter within black holes. General Relativity predicts that the gravitational contraction of a massive star to within the Schwarzschild radius should end with a central singularity. However, General Relativity is incomplete, and in the absence of a theory of *quantum gravity* we have to recognise that we know nothing about the laws which govern the behaviour of matter within black holes. The expanding–contracting black hole Universe tends to suggest that inside a black hole gravitational collapse could be halted before the singularity. A last resistance of matter – for example a highly repulsive interaction which only manifests itself at very small distances – could cause the matter of a collapsed star to 'rebound', and like the whole Universe begin to oscillate indefinitely between

[9] The event horizon should not be confused with the particle horizon, which at a given instant is the frontier of space surrounding the portion of the observable Universe; see 'Géométries de la variété Univers' mentioned in the footnote on page 297.

[10] In fact the maximum radius of the closed Universe is strictly equal to its Schwarzschild radius as measured by an external observer.

a hyperdense state and an expanded state filling the interior of the Schwarzschild sphere. Such behaviour might one day appear in a unified theory of all fundamental interactions, from which gravitational singularities have been eliminated (see Chapter 12).

The black hole universe theory finally raises the question of the uniqueness of our Universe. What would be the status of our closed world with respect to the exterior cosmos? There could be a hierarchy of universes – black holes all situated within one another. Recent theories in physics allow for the existence of such 'bubble-universes'.

These somewhat extravagant speculations belong more to dreams than reality; they do not figure greatly in the work carried out at research institutes because they stretch too far beyond our actual knowledge, and do little for real progress in science. One day perhaps we shall have the tools to help us answer these questions, but we should not delude ourselves: all these theories are based on imagination, and reality is often very different from what we imagine. To capture even a fragment of the real world, we have to work with our brains and our hands, make a thousand measurements, distrust received ideas and theories that are too beautiful.

On reaching the last lines of this book, have we learned anything? I think so. The advent of black holes without doubt marks the beginning of a revolution. A revolution in the changing world of ideas and theories, a revolution in the real world where the fate of stars, galaxies and the Universe itself is slowly being revealed. But all revolutions have their pitfalls. To paraphrase an aphorism of Maurice Maeterlink, the expression *black hole* is often still no more than a sumptuous disguise for our ignorance.

APPENDIX 1
THE HERTZSPRUNG–RUSSELL DIAGRAM

At the beginning of the century the Danish astronomer Ejnar Hertzsprung and the American Henry Russell of Princeton University independently produced a diagram showing the relationship between the luminosity and the surface temperature of stars (Figure A1). Each of the temperature intervals defines a *spectral type*, ranging from O for the hot 'blue' stars to M for the coolest 'red' stars. The Sun is a class G star, 'yellow' of surface temperature 6000 K.

The points on the diagram are not distributed at random; their position reflects the outlines of stellar evolution. The majority of stars are grouped together in a narrow diagonal band, called the *Main Sequence*. This state corresponds to stable hydrogen burning at the centres of the stars. This group includes red dwarfs of small radii and luminosities (the radius is given in brackets in solar units) and very bright blue giants.

Another group of stars extends horizontally above the Main Sequence. This consists of very luminous stars with low temperatures, that is, red giants and supergiants. Stars with low luminosity but high temperatures occupy a region below the Main Sequence: these are stars which have collapsed into white dwarfs.

During its life an individual star moves across the Hertzsprung-Russell diagram. The Sun's evolution is illustrated schematically. The initial contraction phase brings it to the Main Sequence, where it will spend most of its life. When the hydrogen in its core has been used up, the Sun will expand and become a red giant, increasing its radius by a factor of 100 and its luminosity by 1000. Then it will enter a phase of instability, during which it will pulsate with variable luminosity and gradually contract, increasing its surface temperature. Its final destiny, after the ejection of gas in a planetary nebula, will be a slow death as a shrivelled-up white dwarf.

A star 20 times more massive than the Sun will follow a different path, burning more rapidly in its Main Sequence phase; it will expand into a red supergiant and then explode as a supernova to form a neutron star or a black hole, which emit little or no light and are therefore not shown on this diagram.

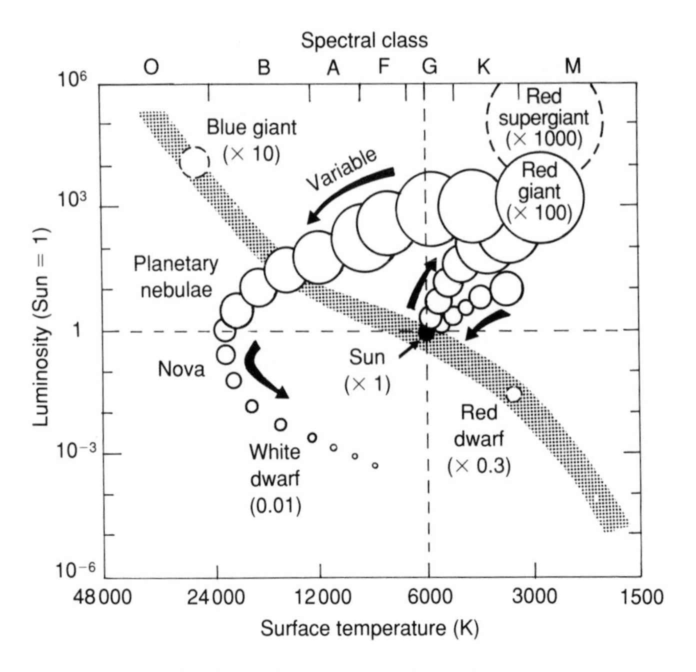

Figure A1. The Hertzsprung–Russell diagram.

APPENDIX 2
THE MASS–DENSITY DIAGRAM OF ASTRONOMICAL OBJECTS AND THE ENDPOINTS OF STELLAR EVOLUTION

A celestial body maintains its equilibrium under the action of opposing forces of compression and expansion. The compression forces may be the electrostatic attraction between the electrons and protons making up atoms and molecules, or gravity, which being attractive always tends to compress. In 'hot' bodies the expansion forces result from thermal pressure because the central temperature is very high. In 'cold' bodies the expansion forces arise from the Exclusion Principle of quantum mechanics, which holds the electron or neutron density above a certain level.

Each equilibrium state is characterised by a relation between its mass and its average density, according to which one or another opposing force comes into play. In Figure A2 the masses and densities are related to the solar values (2×10^{33} grams and 1 g/cm^3), so the Sun is at the origin.

Cold bodies

Cold stars, supported by quantum-mechanical pressure, occupy the black regions. The grey regions are forbidden because they correspond to a violation of the Exclusion Principle.

For bodies less massive than 10^{-3} $M_\odot$, the dominant compression force is electrostatic attraction. The equilibrium state is that of a *planet*, characterised by a density independent of mass and equal to that of normal matter (1 g/cm^3). The point *P* marks the limit of planetary stability, and corresponds approximately to the mass of Jupiter. Above this limit gravitation becomes the dominant compression force and gives rise to various much denser cold equilibrium states.

In *white dwarfs* the internal quantum-mechanical pressure is from degenerate electrons. The density can reach 1 tonne/cm^3. The point *C* is the Chandrasekhar limit, that is, the maximum mass of a white dwarf, 1.4 $M_\odot$. Beyond this limit the electrons become 'relativistic', that is, they acquire velocities close to that of light and are no longer able to support the white dwarf.

In *neutron stars* the internal pressure comes from degenerate neutrons. The matter is much more concentrated, reaching the density of atomic nuclei, 10^{15} g/cm^3. The point *E* marks the limit of stability of neutron stars, at about 3 $M_\odot$. Beyond this limit the neutrons become relativistic and are incapable of supporting the star. *There is no cold equilibrium state for bodies more massive than 3 $M_\odot$.*

Black holes

Black holes occupy the diagonal line which cuts the density axis at *E* and the mass axis at *L*. This latter point corresponds to the black hole envisaged by Michell and Laplace: 10^7 $M_\odot$, 1 g/cm^3. Because gravitation alone governs the state of a black hole, black holes of all masses and densities can in principle exist. *Mini black holes* (bottom of the diagram) are not very massive and are extremely dense; *supermassive black holes* (top of diagram) are by contrast not very dense at all. If the line was extended to a mass of 10^{23} $M_\odot$, a density of 10^{-29} g/cm^3 would result, of the same order as that of the Universe. This suggests that the Universe itself could be the largest black hole of all.

Hot bodies

Hot stars occupy the white region. The Sun and Main Sequence stars are found in a narrow broken band, called a *thermonuclear isotherm* because it corresponds to the central temperature of 10^7 K required to fuse hydrogen into helium. The masses of these stars lie between 0.01 and 100 $M_\odot$.

Stellar evolution

During its evolution a star moves on the mass–density diagram, remaining within the dotted area. Figure A3 is an enlarged version of this region showing the dominant thermonuclear reactions taking place at the centre of the star during its various stages of evolution.

The general tendency of stellar evolution, controlled by gravitation, is an increase in density (motion down the diagram), while phenomena such as mass loss, fragmentation, instability or explosion decrease the mass (motion to the left of the diagram). The evolution of a hot star must end at one of the three possible cold states: white dwarf, neutron star or black hole.

A star less massive than about 8 $M_\odot$ follows curve *A*; after leaving the Main Sequence where hydrogen is transformed into helium, the star's central temperature and density increase until the helium can be turned into carbon. The carbon remains inert and the star finally becomes a white dwarf. Trajectory *B* is that of a more massive star, which succeeds in

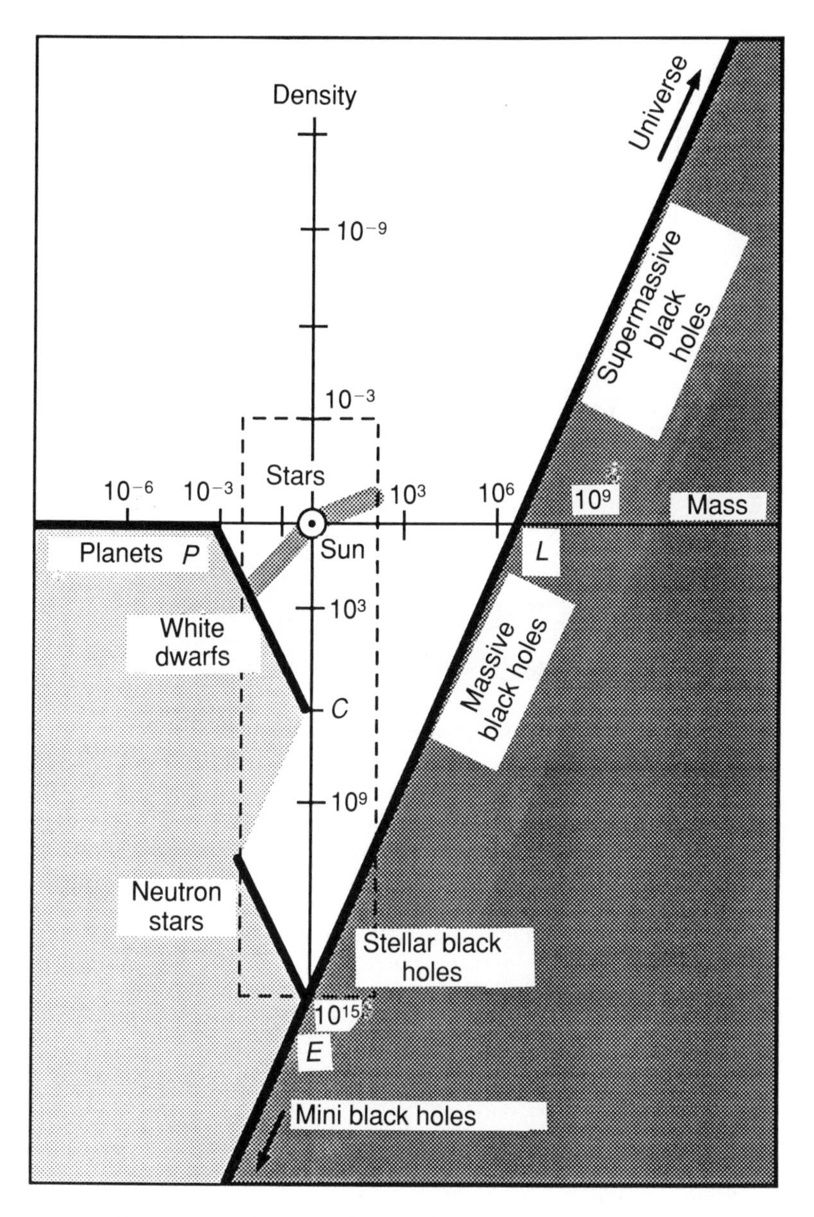

Figure A2. The mass–density diagram of astronomical objects. (From B. Carter.)

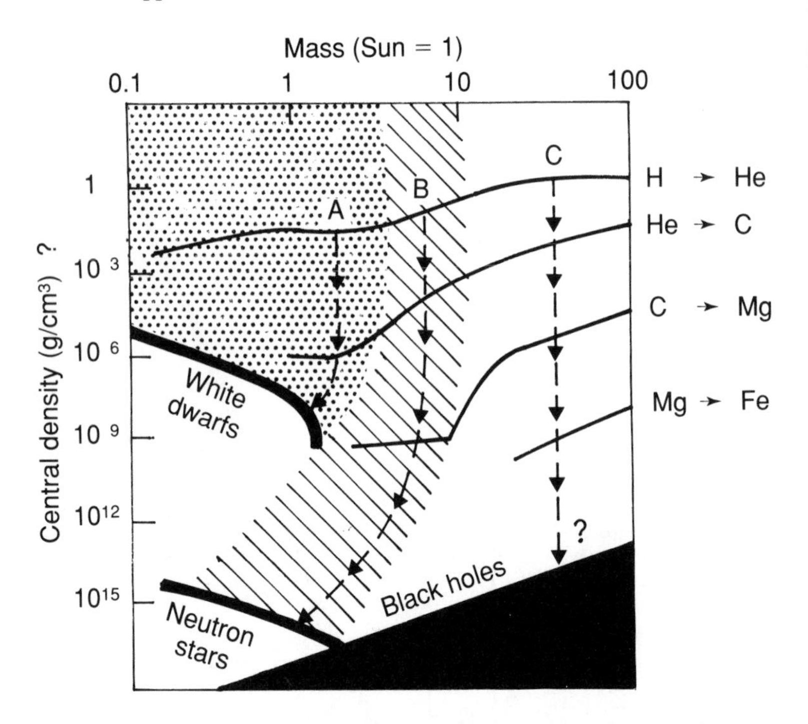

Figure A3. The endpoints of stellar evolution. (From B.Carter.)

burning carbon into magnesium during its evolution. It finally becomes a neutron star. Trajectory *C* is the most hypothetical; it could represent that of a more massive star – more than 25 $M_\odot$ – which, after passing through all the stages of thermonuclear burning to iron, finally becomes a black hole.

Bibliography

Audouze, J. (1981). *Aujourd'hui l'Univers*. Belfond.

Audouze, J. & Israel, G. (eds.) (1988). *The Cambridge Atlas of Astronomy*. Cambridge University Press

Collectif (1987). *Aux Confins de l'Univers*. La Nouvelle Encyclopédie des Sciences et Techniques, Fayard/Fondation Diderot.

Einstein, A. (1954). *Relativity*, 15th edn. Methuen

Einstein, A. & Infeld, L. (1971). *Evolution of Physics: The Growth of Ideas from Early Concepts to Relativity and Quanta*. Cambridge University Press.

Greenstein, G. (1984). *Frozen Star*. Macdonald.

Hawking, S. W. (1988). *A Brief History of Time: From the Big Bang to Black Holes*. Bantam Books.

Hawking, S. W. & Israel, W. (eds.) (1987). *300 Years of Gravitation*. Cambridge University Press.

Heidmann, J. (1989). *Cosmic Odyssey*. Cambridge University Press.

Islam, J. (1983). *The Ultimate Fate of the Universe*. Cambridge University Press.

Kaufmann, W. Jr (1979). *Black Holes and Warped Space Time*. W. H. Freeman.

Misner, C. W., Thorne, K. S. & Wheeler, J. A. (1973). *Gravitation*. W. H. Freeman.

Montmerle, T. & Prantzos, N. (1988). *Soleils Éclatés*. Presses du CNRS.

Nicolson, I. (1981). *Gravity, Black Holes and the Universe*. John Wiley & Sons.

Reeves, H. (1984). *Atoms of Silence: An Exploration of Cosmic Evolution*. MIT Press.

Ruffini, R. & Wheeler, J. A. (1971). Introducing the black hole, *Physics Today*, January 1971

Sullivan, W. (1979). *Black Holes: The Edge of Space, the End of Time*. Anchor Press/Doubleday.

Wheeler, J. A. (1990). *A Journey into Gravity and Spacetime*. Scientific American Library.

Will, C. (1986). *Was Einstein Right?* Basic Books.

Name index

Subject index